MICROWAVE FREQUENCY SYNTHESIZERS

MICROWAVE FREQUENCY SYNTHESIZERS

RONALD C. STIRLING
TECHNICAL STAFF MEMBER
SCIENTIFIC-ATLANTA
ELECTRO PRODUCTS DIVISION

PRENTICE-HALL, INC.
Englewood Cliffs, New Jersey 07632

Library of Congress Cataloging-in-Publication Data

Stirling, Ronald C., 1931–
Microwave frequency synthesizers.

Includes bibliographies and index.
1. Frequency synthesizers. 2. Microwave devices.
I. Title.
TK7872.F73S75 1987 621.3815′363 86–91457
ISBN 0–13–582016–2

Editorial/production supervision and
interior design: Richard Woods
Cover design: Wanda Lubelska
Manufacturing buyer: Gordon Osbourne

Printed in the United States of America

10 9 8 7 6 5 4 3 2 1

ISBN 0-13-582016-2 025

Prentice-Hall International (UK) Limited, *London*
Prentice-Hall of Australia Pty, Limited, *Sydney*
Prentice-Hall Canada Inc., *Toronto*
Prentice-Hall Hispanoamericana, S.A. *Mexico*
Prentice-Hall of India Private Limited, *New Delhi*
Prentice-Hall of Japan, Inc., *Tokyo*
Prentice-Hall of Southeast Asia Pte. Ltd., *Singapore*
Editora Prentice-Hall do Brasil, Ltda., *Rio de Janeiro*

To my wife, Coleen,
for her support and encouragement to write this book.

Contents

Preface

The objectives of this book are to provide the practicing engineer with a reference for the design of wideband microwave frequency synthesizers with emphasis on the capability to switch between frequencies in fractions of a microsecond. To meet these objectives, numerous examples of different design techniques are described to emphasize the fundamental limits imposed on the synthesizer switching time by the circuits used in each of the various technical approaches. In addition to these objectives, design criteria are established to achieve low phase noise and reduction of spurious responses. The application of this book for the purpose of selecting a particular synthesizer design approach should be useful to the engineer designing frequency synthesizers in the test equipment, radar, and communications fields.

The organization of the book is to begin each major classification of frequency synthesizer type (direct, indirect, and combination) with a discussion of the various required circuit functions and then combine these functions to make a rudimentary synthesizer design. The basic technical approaches are then developed from these simpler designs to more complex designs containing large bandwidths in the microwave frequency bands.

Initially, in the first chapter, definitions of single-sideband phase noise are developed in accordance with present usage. In addition, the methods and test equipment required to measure sideband phase noise are described. The first major classification of synthesizers discussed is the direct type, in Chapters 2 through 5. This section begins with a discussion of the circuits common to direct-type techniques and how they are combined into a synthesizer design. The various designs described use base 2, 4, 8, and 10 for the frequency synthesis techniques.

The second major classification is the indirect-type, in which phase-locked loops are employed to synchronize an electronically tuned oscillator to single- and multiple-reference frequencies, discussed in Chapters 6 through 13. In the first of these chapters, circuits needed in these designs are described and then combined to make the elementary single-loop synthesizer. The designs are expanded to more complex wideband frequency synthesizers. In addition, a new design approach is introduced, using programmable frequency dividers to provide a variable reference to the synthesizer's phase detector.

Chapters 14 through 18 discuss the advantages of using combined direct and indirect techniques to optimize performance versus complexity as a function of slower frequency switching speeds. In addition, methods to add the capability of frequency modulation and phase modulation to wideband microwave frequency synthesizers are described.

In recent times, linear monolithic circuits fabricated from silicon and GaAs materials have shown the capability to make the direct-type design smaller and lighter by using this technology to make future synthesizer circuits from these materials. For these reasons, the direct-type synthesizer can be a small and lightweight realizable design where fractional microsecond switching speeds are necessary to meet critical system objectives.

Acknowledgments

All of the material presented in this book is based on a wide variety of assignments with my former employer, Sperry Microwave Electronics in Clearwater, Florida. Several former engineering colleagues at Sperry contributed to numerous frequency synthesizer designs for a wide variety of applications; to them, I am grateful for all of our discussions related to the design of frequency synthesizers.

In particular, I would like to thank engineers John P. Jones, John Kettle, Dr. James Adair, José Alonso, and Robert Thurmond. I would also like to thank my current employer, Scientific-Atlanta, for the use of their facilities to improve this book's manuscripts. I am also grateful to Mrs. Jeannie Harper of Tullahoma, Tennessee, for her efforts to make the original manuscript meet the publisher's requirements.

R. C. S.

1

Introduction

The first part of this chapter is a brief review describing the "spectral purity" for a CW source of RF energy. Definitions are established based on frequently used equations for describing phase noise of RF sources. From these definitions, expressions for power spectral density and equivalent frequency deviation are derived. The chapter concludes with a review of the test fixtures containing frequency discriminators and phase detectors, used to measure frequency deviation and phase noise. At the end of this chapter, six references are listed to provide the reader with a more comprehensive analysis on the theory and measurement of phase noise.

Chapters 2 through 5 describe direct-type frequency synthesizer design techniques. Chapter 2 is a review of circuits used in direct synthesis. The remaining three chapters examine basic concepts and design techniques used in the direct-type designs, with emphasis on obtaining fractional microsecond frequency set-on time, the time required to switch from one frequency to another.

Chapters 6 through 13 describe indirect-type frequency synthesizer design techniques. Chapters 6 and 7 discuss phase detectors and phase-locked loops, including a description of feedback loop stability and its relation to an oscillator's closed-loop spectral purity. In Chapter 8 these concepts are expanded to include digital phase-locked loops. Chapters 9 and 10 expand the single digital loop analysis to multiple loops and conclude with their application to wideband microwave frequency synthesizers. Design examples are described in Chapter 11, and several novel design techniques are presented in Chapter 12. Chapter 13 presents a description of techniques to measure the synthesizer's electrical performance.

Chapters 14 through 18 are concerned with combining direct and indirect techniques to achieve either fast frequency set-on times or lower cost in slower switching speed requirements. In Chapters 17 and 18, phase and frequency modulation is added to the synthesizer using combined indirect and direct techniques.

SPECTRAL PURITY OF RF SOURCES

The instantaneous voltage from a CW source of RF energy without amplitude or angular modulation can be expressed as

$$V(t) = V_p \sin \omega_c t \tag{1-1}$$

where V_p is the peak voltage and $\omega_c t$ is the linear increasing phase. Generally, at offset frequencies close to the carrier frequency f_c, where $\omega_c = 2\pi f_c$, the noise power from randomly caused amplitude variations in V_p are much smaller than the random phase variations about the linear progressing phase and will be neglected in this discussion.

For an RF source with sinusoidal phase modulation, the waveform can be expressed as:

$$V(t) = V_p \sin\left(\omega_c t + \frac{1}{Y}\sin \omega_m t\right) \tag{1-2}$$

where ω_m is the radian modulation frequency ($\omega_m = 2\pi f_m$) and $1/Y$ is the peak radian ($\Delta\phi_p$) phase deviation and is the modulation index. For the condition of $Y >> 1$, Eq. (1-2) can be expanded as follows:

$$V(t) = V_p \left\{(\sin \omega_c t)\left[\cos\left(\frac{1}{Y}\sin \omega_m t\right)\right] + \cos \omega_c t\left[\sin\left(\frac{1}{Y}\sin \omega_m t\right)\right]\right\} \tag{1-3}$$

Note:

$$\cos\left(\frac{1}{Y}\sin\omega_m t\right) \approx 1 \quad \text{for } Y >> 1$$

and

$$\sin\left(\frac{1}{Y}\sin\omega_m t\right) \approx \frac{1}{Y}\sin\omega_m t \quad \text{for } Y >> 1$$

therefore,

$$V(t) = V_p\left(\sin\omega_c t + \frac{1}{Y}\cos\omega_c t \sin\omega_m t\right) \tag{1-4}$$

The term $\frac{1}{Y}\cos\omega_c t \sin\omega_m t$ can be expanded to

$$\frac{1}{2Y}[\sin(\omega_c t + \omega_m t) + \sin(\omega_c t - \omega_m t)]\ V(t)$$

$$= V_p\left[\sin\omega_c t + \frac{1}{2Y}\sin(\omega_c t + \omega_m t) + \frac{1}{2Y}\sin(\omega_c t - \omega_m t)\right] \tag{1-5}$$

In the frequency domain, Eq. (1-5) represents a carrier frequency f_c with sidebands at $\pm f_m$ from the carrier and sideband voltage amplitudes equal to $V_p/2Y$. The ratio of either sideband's voltage to the carrier's voltage is

$$\frac{V_{\text{ssb}}}{V_c} = \frac{1}{2Y} \tag{1-6}$$

and expressed as a power ratio in dB:

$$\left.\frac{P_{\text{ssb}}}{P_c}\right|_{\text{dB}} = 20\log\frac{1}{2Y} \tag{1-7}$$

If we consider the random phase noise in an RF source to be composed of a large number of uncorrelated perturbations spaced 1 Hz apart, each with 1 Hz of bandwidth, we can use Eqs. (1-2) through (1-5) to describe the phase and noise voltage deviations at a frequency removed from the carrier by f_m.

When the phase noise is observed on a spectrum analyzer, displayed on a dB power scale, Eq. (1-7) can be used to measure 20 log $1/2Y$, and we can compute $1/2Y$. This quantity is one-half of the peak deviation $\Delta\phi_p$ and is equivalent to $\Delta\phi_{\text{rms}}/\sqrt{2}$, where $\Delta\phi_{\text{rms}}$ is the rms radian phase deviation.

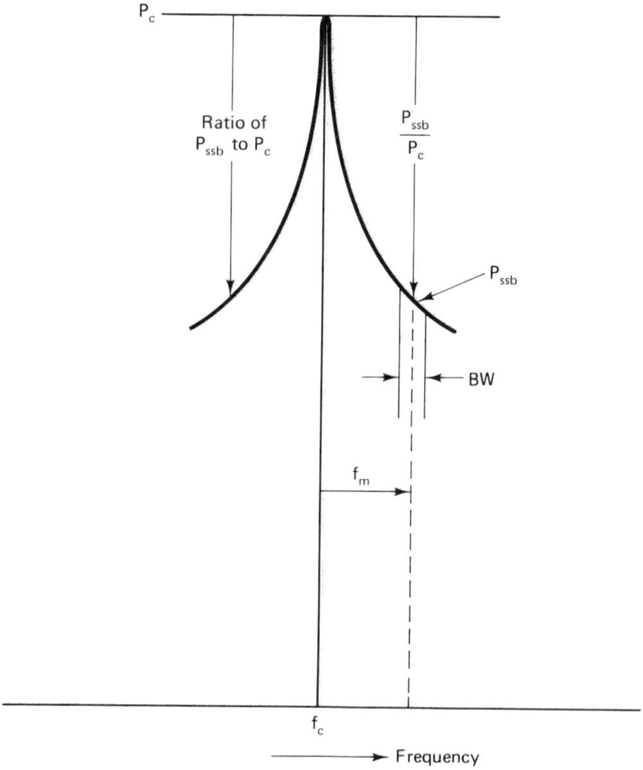

Figure 1-1. Ratio of single-sideband to carrier power

Returning to Eq. (1-6),

$$\frac{V_{\text{ssb}}}{V_c} = \frac{1}{2Y} = \frac{\Delta\phi_p}{2} \tag{1-6}$$

The ratio squared of the single-sideband voltage to carrier voltage in a 1-Hz bandwidth at a frequency removed (f_m) from the carrier power is a power ratio of sideband noise to carrier power and is represented by $\mathcal{L}(f_m)$ as defined by the U.S. National Bureau of Standards. It can be derived from a power spectrum as illustrated in Fig. 1-1 when the measured sideband phase noise power is converted from the measurement bandwidth to 1 Hz. $\mathcal{L}(f_m)$ is normally illustrated as a function of f_m, logarithmically, expressed in dB below the carrier. The single-sideband power to carrier power ratio is defined as

$$\frac{P_{\text{ssb}}}{P_c} = \left(\frac{V_{\text{ssb}}}{V_c}\right)^2 = \left(\frac{1}{2Y}\right)^2 = \frac{1}{4}\left(\frac{1}{Y}\right)^2 = \frac{\Delta\phi_p^2}{4} \tag{1-8}$$

recalling that $\Delta\phi_p$ is the modulation index, $\mathcal{L}(f_m)$ is simply the square of the phase modulation index divided by 4. When a phase detector is used to measure the rms voltage output and is related to the rms phase noise by a constant, it is more convenient to represent Eq. (1-8) by

$$\frac{P_{\text{ssb}}}{P_c} = \frac{\Delta\phi_{\text{rms}}^2}{2} \tag{1-9}$$

To summarize, the relationship between power ratios and modulation index has been established in Eqs. (1-6) through (1-9).

Discriminators are used occasionally to measure the sideband noise of a carrier, and under these conditions the quantity being measured is frequency deviation. We can relate the frequency deviation to phase noise by taking the derivative of the instantaneous phase of Eq. (1-2) as follows:

$$\text{Frequency deviation} = \frac{d\phi}{dt} = \frac{d\left(\omega_c t + \dfrac{1}{Y}\sin\omega_m t\right)}{dt} \tag{1-10}$$

$$\frac{d\phi}{dt} = \omega_c + \frac{\omega_m}{Y}\cos\omega_m t \equiv \text{frequency}$$

The frequency term is composed of the constant carrier at f_c and a peak deviation of f_m/Y at a modulation rate of f_m. Since

$$\Delta\phi_p = \frac{1}{Y} \tag{1-11}$$

the peak deviation of f_m/Y is

$$\Delta f_p = \frac{f_m}{Y} = f_m\,\Delta\phi_p \tag{1-12}$$

$$\frac{\Delta f_p}{f_m} = \Delta\phi_p = \frac{1}{Y} = \text{modulation index} \tag{1-13}$$

If we multiply both sides by ½, we obtain

$$\frac{\Delta f_p}{2f_m} = \frac{1}{2Y} = \frac{V_{\text{ssb}}}{V_c} \quad \text{from Eq. (1-6)} \tag{1-14}$$

Therefore, the equivalent peak frequency deviation is directly related to the sideband to carrier voltage ratio. If we square all sides of Eq. (1-14), we have

$$\left(\frac{V_{\text{ssb}}}{V_c}\right)^2 = \left(\frac{\Delta f_p}{2f_m}\right)^2 = \left(\frac{\Delta f_{\text{rms}}}{\sqrt{2}\,f_m}\right)^2 \tag{1-15}$$

and we can state the relationship of Eqs. (1-8), (1-9), (1-14), and (1-15) as

$$\frac{P_{\text{ssb}}}{P_c} = \left(\frac{V_{\text{ssb}}}{V_c}\right)^2 = \left(\frac{\Delta\phi_{\text{rms}}}{\sqrt{2}}\right)^2 = \left(\frac{\Delta f_{\text{rms}}}{\sqrt{2}\, f_m}\right)^2 \tag{1-16}$$

Taking the log and multiplying by 10:

$$\left.\frac{P_{\text{ssb}}}{P_c}\right|_{\text{dB}} = 20 \log \frac{V_{\text{ssb}}}{V_c} = 20 \log \frac{\phi_{\text{rms}}}{\sqrt{2}} = 20 \log \frac{\Delta f_{\text{rms}}}{\sqrt{2}\, f_m} \tag{1-17}$$

In Eqs. (1-16) and (1-17), we have established the relationships between power ratios, phase deviations, and frequency deviations as a function of the frequency removed (Fourier frequency) in the frequency domain. We shall use these equations in the following discussions on how these terms are measured.

MEASURING SPECTRAL PURITY

Frequency synthesizers usually have a low-frequency (5 to 10 MHz), high-stability standard that determines the long-term frequency stability of the synthesizer's output. For this type of synthesizer, the combined phase noise of two frequency synthesizers can be measured as diagrammed in Fig. 1-2. With the two frequency synthesizers at the same frequency, the measured rms voltage at the frequency-selective voltmeter's input contains the sum of both synthesizers' phase noise and is a double-sideband measurement. If the synthesizers' sidebands contribute equally to the measured value, the single-sideband amplitude of either synthesizer is equal to one-half the measured value.

The measurement begins with calibrating the test fixture by the insertion of a voltage-controlled phase modulator placed between *A* and *B* at the output of Synthesizer 1. The modulator is used to phase-modulate this synthesizer at a low modulation index with a modulation frequency of f_m. The amount of modulation is observed on the spectrum analyzer and is adjusted to be a specific value in the range of 20 to 40 dB below the carrier frequency f_c. Next the variable phase shifter is adjusted for a maximum voltage indication. For each specific value of modulation frequency, the single-sideband amplitude is set to a specific value (30 dB down, for example), and the measured voltage is recorded.

The phase detector is removed and *A* is reconnected to *B*. At the lowest value of recorded modulation frequency, the variable phase shifter

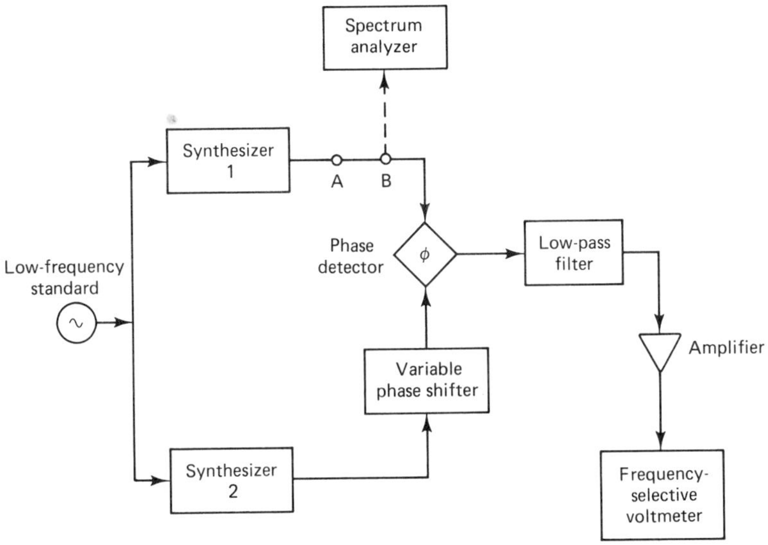

Figure 1-2. Two-source method of measuring phase noise

is again adjusted for a maximum voltage measured at the voltmeter. For the remaining values of recorded f_m, the selective voltmeter is tuned to each one of the previously recorded modulation frequencies, and the voltage is measured.

For the phase detector, the phase deviation is given by

$$\phi_{\text{rms}} = K_\phi V_{\text{rms}} \tag{1-18}$$

Since we have a measurement of V_{rms} in the calibrate mode and a corresponding value of the P_{ssb}/P_c ratio, Eqs. (1-16) and (1-17) can be used to calculate ϕ_{rms}. Knowing the value of ϕ_{rms} allows us to calculate the phase detector constant K_ϕ. With K_ϕ a known value, we can convert the measured noise voltages to an rms phase deviation at each value of f_m.

Occasionally, only one synthesizer is available, and this will require a measurement configuration as illustrated in Fig. 1-3. As in the previous example, the setup is also a double-sideband measurement and must be converted to a single-sideband measurement by dividing the measured rms voltage by $\sqrt{2}$. The configuration is calibrated initially by replacing the synthesizer with a voltage-controlled oscillator (VCO) at the same frequency and power output as the synthesizer. The VCO is modulated at a modulation frequency f_m, and the modulation index is limited to a value much less than unity by adjusting the sidebands to be 30 dB down as measured on the spectrum analyzer. At each modulating frequency, the sidebands are set at -30 dBc, and the voltage is recorded.

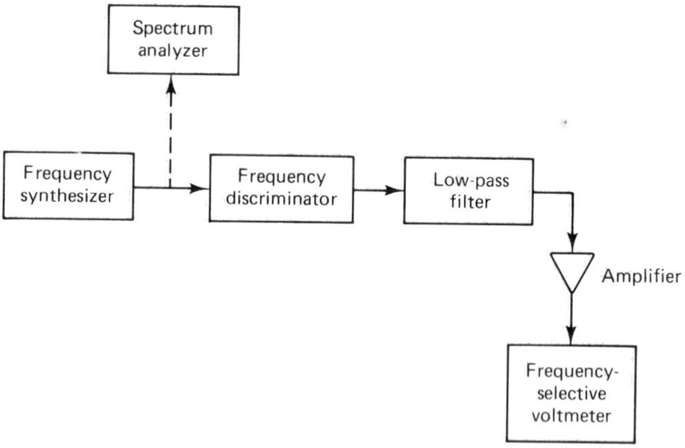

Figure 1-3. Single-source method of measuring phase noise

The VCO is removed and replaced with the synthesizer to be measured. The selective voltmeter is tuned to the previously recorded values of f_m, and the voltage is measured. To convert the measured noise voltages to the P_{ssb}-to-P_c ratio defined by Eq. (1-17), it is necessary to determine the discriminator constant defined for

$$\Delta f_{rms} = K_d V_{rms} \tag{1-19}$$

In the calibrate mode, a specific value of Δf_{rms} can be calculated for the ratio of P_{ssb} to P_c at -30 dBc. Once Δf_{rms} has been computed, the value of K_d can be computed from Eq. (1-19). To compute the noise-to-carrier ratio, the measured noise voltage is used to compute the equivalent noise deviation Δf_{rms}, and Eq. (1-17) can then be used to calculate the single-sideband to carrier noise ratio.

REFERENCES

1. Allan, D. W., "Statistics of Atomic Frequency Standards," *Proceedings of the IEEE*, 54, no. 2 (February 1966), 221–30.
2. Cutler, L. S. and Searle, C. L., "Some Aspects of the Theory and Measurement of Frequency Fluctuations in Frequency Standards," *Proceedings of the IEEE*, 54, no. 2 (February 1966), 136–54.
3. Barber, R. E., "Short Term Frequency Stability of Precision Oscillators and Frequency Generators," *The Bell System Technical Journal*, vol. 50, no. 3 (March 1971).

4. Barnes, J. A., Chi, A. R., Cutler, L. S., Healey, D. J., Leeson, D. B., McGunigal, T. E., Mullen, J. A. Jr., Smith, W. L., Sydnor, R. L., Vessot, R. F. C., Winkler, G. M. R., "Characterization of Frequency Stability," *IEEE Transactions on Instrumentation and Measurement*, IM-20, no. 2 (May 1971), 105–20.

5. Allan, D. W., "The Measurements of Frequency and Frequency Stability of Precision Oscillators," *NBS Technical Note* 669 (May 1975).

6. Howe, D. A., "Frequency Domain Stability Measurements: A Tutorial Introduction," *NBS Technical Note* 679 (March 1976).

2

Circuits Used in Direct-Frequency Synthesizers

In this and the next three chapters, direct-frequency synthesizers' design techniques are described. This chapter is a review of circuits essential for direct-type designs. Chapter 3 introduces the basic concepts employed in direct synthesizers and performance limits imposed by internal circuits. Chapter 4 is a description of design techniques commonly used in past and current designs and ends with a brief discussion on the relative merits of each technique. Chapter 5 ends the discussion on direct-type synthesizers with design examples and measured performance of synthesizers that the author and numerous associates have designed.

ARITHMETIC FUNCTIONS

Circuits employed in direct-type synthesizers provide the arithmetic functions of addition, subtraction, multiplication, and division that are required in this type of synthesizer. In nearly all applications, frequency mixers are commonly used to generate the sum or difference of two periodic waveforms with unequal sinewave frequencies. Multiplication of an RF signal's frequency is obtained by applying the sinewave to a nonlinear device that causes harmonics of the input signal frequency to be generated. Frequency division is normally obtained with digital counters, which may be either fixed or programmable. Unfortunately, all of these devices produce unwanted spurious frequencies that must be suppressed. The output of a typical mixer contains, at reduced power levels, a wide spectrum of signals, including each input frequency and its harmonic intermodulation products. Frequency divider outputs may have a pulsed waveform or a square wave; these waveforms contain harmonics. Frequency multipliers also contain many harmonically related frequencies that need to be attenuated. Because of these spectral purity problems, filters are always placed at the output of circuits providing the four basic arithmetic functions.

FILTERS

Filters used for this equipment are usually low-pass or narrow-bandpass types. The bandpass filter is normally placed at a mixer's output to select either the sum or the difference frequency and attenuate all others outside the passband. Bandpass filters are also placed on a multiplier's output to select the desired harmonic and reject all other harmonics. For frequency dividers, low-pass filters are used to attenuate all harmonics of the pulsed-type waveform and allow transmission of the fundamental frequency component.

The degree to which unwanted spurious signal responses are attenuated will define the number of required filter sections in a bandpass filter with a defined bandwidth. The single most important property of a filter for use in fast switching applications is bandwidth. As bandwidth is increased, the time for a signal to propagate through the filter decreases [1].* This dependence is initially illustrated with the aid of Fig. 2-1, where the switch is assumed to connect the signal ($V_p \cos \omega t$) from the termination to the filter in zero time.

Letting τ_d represent delay time from 0% to 10% of final amplitude and τ_r represent 10% to 90% amplitude risetime measured at the load, we can see the importance of maintaining large bandwidth and keeping the number of filter sections to a minimum, as illustrated in Fig. 2-2. The delay time (0% to 10%) for a Chebyshev filter with 0.01 dB ripple was

*Numbers in brackets refer to similarly numbered references in the list at the end of each chapter.

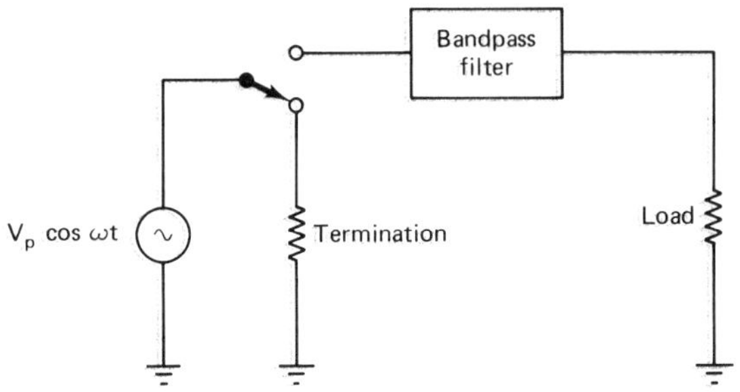

Figure 2-1. RF-signal applied to a bandpass filter

computed from the risetime (10% to 90%) where risetime is approximated as the reciprocal of 3 dB bandwidth in narrow bandpass filters.

The delay time is a function of bandwidth and number of filter sections, decreasing with larger bandwidth and increasing as more sections are added to the filter; therefore, to minimize delay time, the number of sections is held to a minimum, and bandwidth should be as large as possible. Risetime, τ_r, illustrated in Fig. 2-3, is approximately equal to the reciprocal of 3 dB BW for filters when the input signal is at the filter center frequency or near it. From Fig. 2-3 we can see the importance of performing frequency synthesis at microwave frequencies with large-bandwidth signal paths in order to minimize τ_r. It is also important to have input signal frequencies not too far from the filter center frequency in order to prevent "ringing," or excessive delay time (group delay) can result if the signal frequency is at the bandwidth edges.

Group delay characteristics of a filter are useful for approximating the filter bandwidth needed to allow signals in a specific band to be switched through a filter without increasing $\tau_d + \tau_r$ significantly from the center frequency values and causing unwanted damped oscillations. Figure 2-4 shows the most desired region of operation to prevent excessive switching time through the filter and undesired damped oscillations. These relationships between bandwidth and the time a switched signal takes to propagate through a filter can also be applied to any bandwidth-limited device (a tuned amplifier or a frequency multiplier); however, a different type of delay is encountered when we change the input signal frequency to a frequency divider derived from a digital counter.

Since this device lowers the input frequency, based on an accumulated count of input zero crossings, the maximum delay time is simply the reciprocal of the lowest frequency output in the "worst case." Here, we

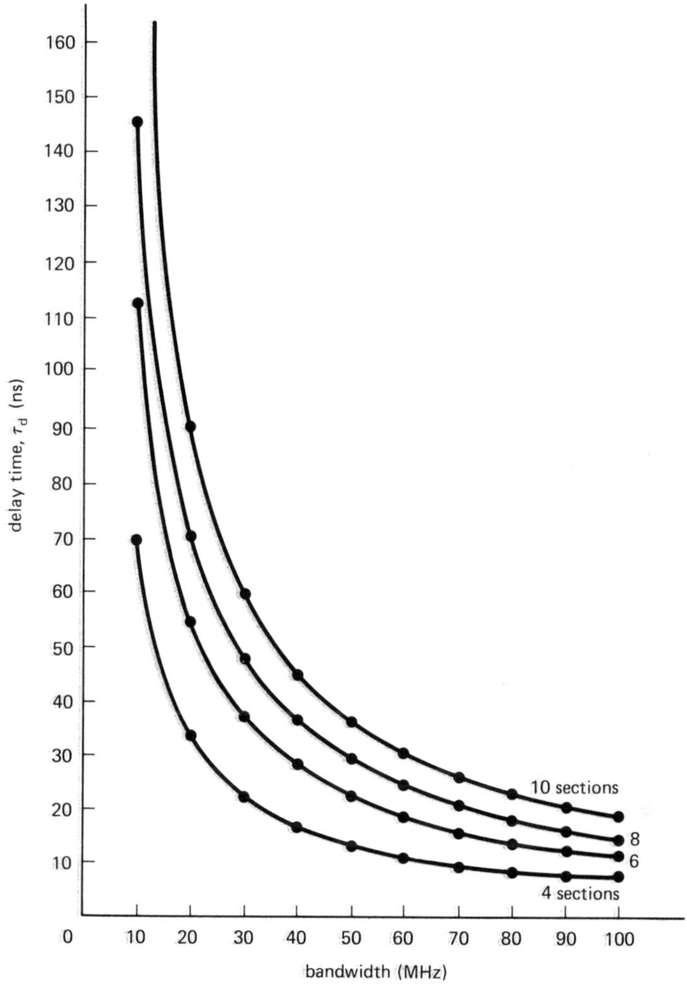

Figure 2-2. Example of delay time as a function of bandwidth and sections. Adapted from A. I. Zverev, *Handbook of Filter Synthesis* (New York: John Wiley & Sons, 1967), p. 403, with permission from the publisher.

define worst case as no accumulated counts within the divider, and this can occur if a new frequency is switched into the divider slightly after the divider's output has changed states. This condition requires the divider to accumulate counts until the output changes states again and is equal to the time period of one cycle of output frequency.

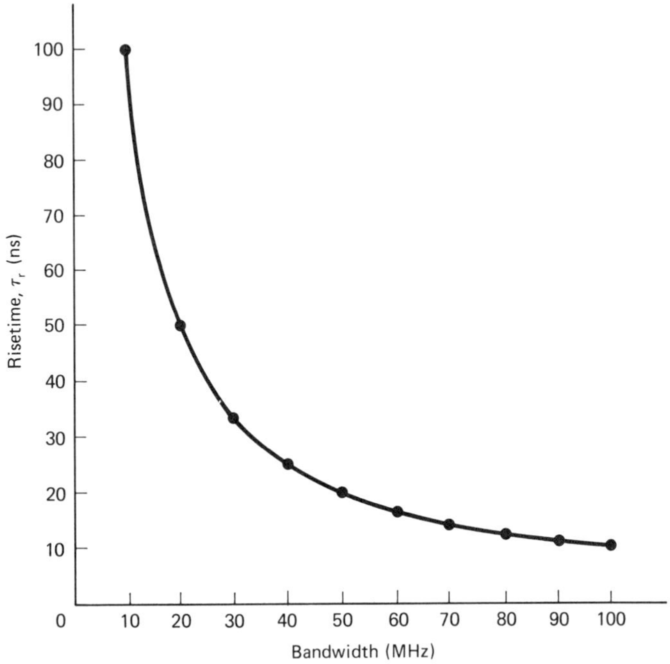

Figure 2-3. Example of risetime as a function of 3 dB bandwidth. Adapted from A. I. Zverev, *Handbook of Filter Synthesis* (New York: John Wiley & Sons, 1967), p. 403, with permission from the publisher.

MIXERS

The unfiltered output of a passive diode mixer is a spectrum of signal frequencies that are related by $f_0 = nf_l \pm mf_r$, where n and m are integers starting at zero, with the higher power input at frequency f_l and the lower power input at frequency f_r [2]. The sums and differences are commonly referred to as harmonic modulation products; fortunately, the power levels of these unwanted spurious responses rapidly decrease to insignificant levels when n and m are greater than 10 in balanced diode mixers.

The ideal single balanced mixer with perfect internal matching and resistive impedances on all ports would have outputs as listed in Table 2-1. From this table we see that even harmonics of the lower-level input frequency f_r are not present in the output with ideal internal matching and resistive impedances. These conditions can be approached at lower frequencies; however, to obtain fast-switching synthesizers, we need to op-

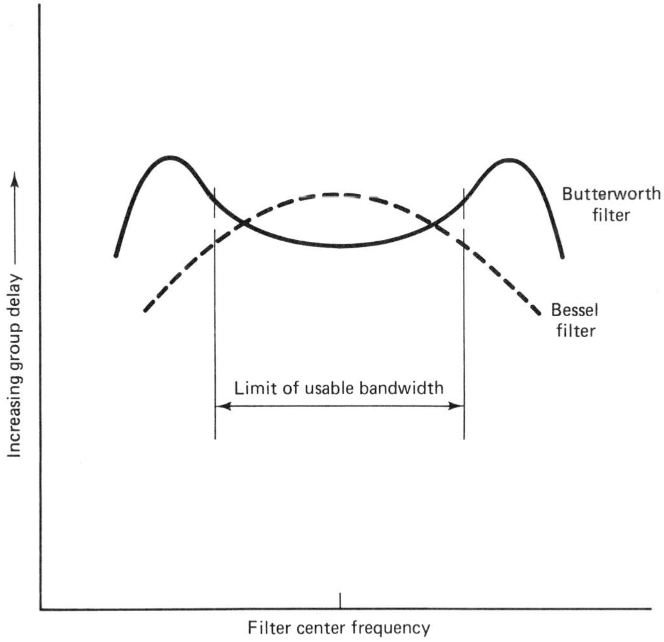

Figure 2-4. Group delay near a filter's center frequency

erate at microwave frequencies, and we need a mixer with the least amount of spurious output signal responses.

The best performance may possibly be obtained with double-balanced mixers that use matched active FETs; however, these are not widely available, and we will limit this discussion to double-balanced diode types.

TABLE 2-1

Harmonics of f_r	Harmonics of f_l					
	0	f_l	$2f_l$	$3f_l$	$4f_l$	$5f_l$
0	0	f_l	$2f_l$	$3f_l$	$4f_l$	$5f_l$
f_r	f_r	$f_l \pm f_r$	$2f_l \pm f_r$	$3f_l \pm f_r$	$4f_l \pm f_r$	$5f_l \pm f_r$
$2f_r$	$2f_r$					
$3f_r$	$3f_r$	$f_l \pm 3f_r$	$2f_l \pm 3f_r$	$3f_l \pm 3f_r$	$4f_l \pm 3f_r$	$5f_l \pm 3f_r$
$4f_r$	$4f_r$					
$5f_r$	$5f_r$	$f_l \pm 5f_r$	$2f_l \pm 5f_r$	$3f_l \pm 5f_r$	$4f_l \pm 5f_r$	$5f_l \pm 5f_r$

Source: R. B. Mouw and S. M. Fukuchi, "Broadband Double Balanced Mixer/Modulators," *Microwave Journal* (March 1969), Fig. 3; reprinted with permission from the publisher.

TABLE 2-2

Harmonics of f_r	Harmonics of f_l					
	0	f_l	$2f_l$	$3f_l$	$4f_l$	$5f_l$
0	0	f_l		$3f_l$		$5f_l$
f_r	f_r	$f_l \pm f_r$		$3f_l \pm f_r$		$5f_l \pm f_r$
$2f_r$						
$3f_r$	$3f_r$	$f_l \pm 3f_r$		$3f_l \pm 3f_r$		$5f_l \pm 3f_r$
$4f_r$						
$5f_r$	$5f_r$	$f_l \pm 5f_r$		$3f_l \pm 5f_r$		$5f_l \pm 5f_r$

Source: R. B. Mouw and S. M. Fukuchi, "Broadband Double Balanced Mixer/Modulators," *Microwave Journal* (March 1969), Fig. 4; reprinted with permission from the publisher.

From Table 2-2 we see that only odd harmonics of the inputs generate spurious output responses in the idealized case and is a better choice than the single-balanced mixer. The best approach to selection of a specific mixer is to consult manufacturers of these devices and obtain an estimate of the spurious power levels when terminated in nonidealized loads for a defined operating signal level and frequency band.

REFERENCES

1. Zverev, A. I., *Handbook of Filter Synthesis*. New York: John Wiley & Sons, 1967.
2. Peter, W. "Reactive Loads: The Big Mixer Menace," *Microwaves* (April 1971), pp. 38–42.

3

Elementary Synthesizer Designs

SYNTHESIS WITH DIVIDERS, FILTERS, AND SWITCHES

One of the highest operating frequency dividers is a divide-by-2 using digital-type circuits. These devices can be used to lower the frequency of a synthesized signal source as shown in Fig. 3-1.

In this example, we let f_s have an operating range from f_{s1} to f_{s2} with a minimum tuning step value of Δf_s. The dividers not only lower the operating band by successive factors of 2, but they also reduce the minimum step size by a factor of 2 at each divider's output.

The circuit arrangement shown in Fig. 3-2 is capable of dividing f_s by either 2, 4, 8, or 10 (and Δf_s by the same amount) from the single input at f_s to a single output at f_o. The bandpass filters are needed to reduce the level of unwanted harmonics present at each divider's output. For the present, assume that the RF switches are set to select the divide-by-2 path and are instantaneously switched to the divide-by-8 path. The output at f_o will go to zero power until the signal propagates through the divide-by-8 in a time period of $1/8\,f_s$ and the delay time through the bandpass filter. Although successive stages of frequency dividers could be used to obtain

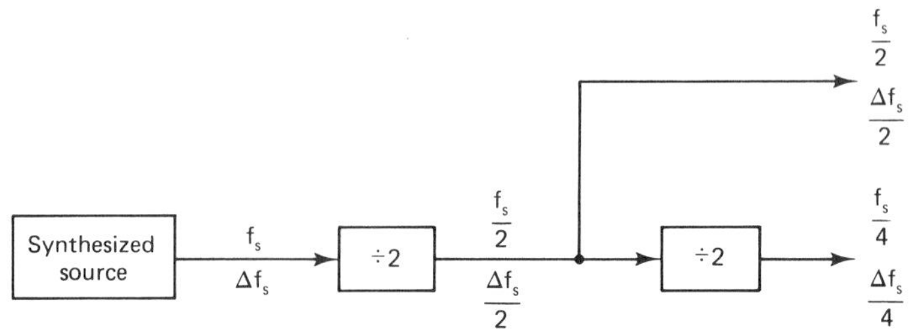

Figure 3-1. Lowering frequency with digital dividers

lower frequencies from a higher available band, it causes slower frequency set-on times as the output frequency decreases.

One other technique to obtain lower frequencies would be to use a programmable divider (N), as shown in Fig. 3-3, with switches and filters added for selection of the appropriate filter to reduce undesired harmonics at the divider's output. The major problem associated with this approach is the limited discrete values of N that will enable f_s/N to remain an integer.

Using frequency dividers as one of our synthesizer building blocks does provide one very useful property besides division of frequency. Spurious sideband power levels are reduced by 20 log N at the divider's output, and this property increases carrier-to-noise ratio of the divided signal frequency. This improvement in spectral purity results because (for low modulation index) peak frequency deviation is divided and frequency deviation rate is not divided; thus the modulation index of noise and spurious sidebands is divided by N and causes an increase in spectral purity.

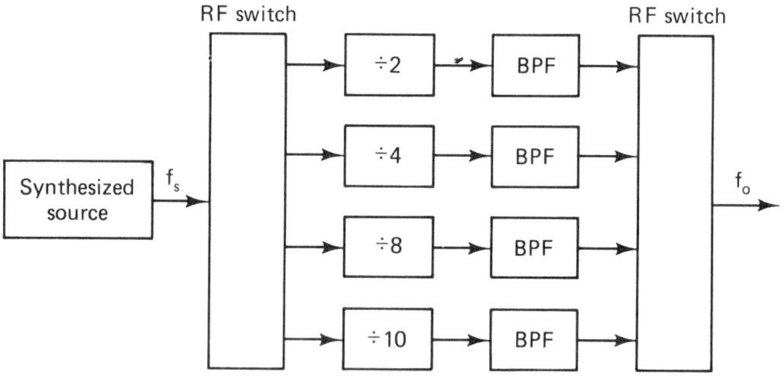

Figure 3-2. Switched dividers and filters

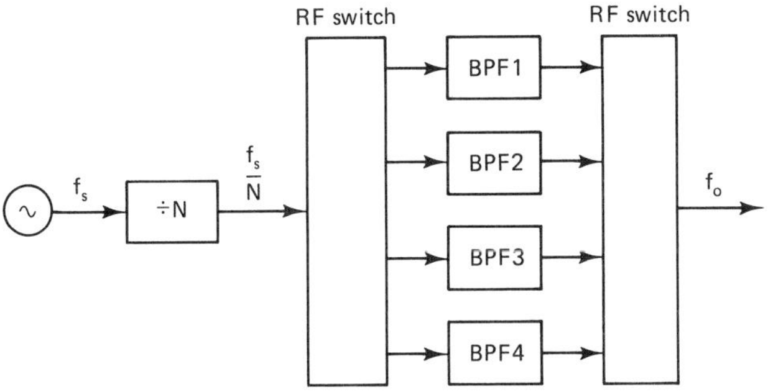

Figure 3-3. Switched bandpass filters

SYNTHESIS WITH MULTIPLIERS, FILTERS, AND SWITCHES

Types of frequency multipliers. Multiplication of frequency is usually obtained by interrupting the periodic waveform in a manner to cause sharp discontinuities; the resultant waveform contains the fundamental frequency component and harmonics of the fundamental. Generally, multipliers derived from resistive devices have the largest conversion power loss, while those derived from reactive devices have the lowest conversion power loss and, unfortunately, the highest potential for instability.

Generally, multipliers derived from resistive devices are wideband and can be made to operate over several octaves, without tuning or bias circuits, at the expense of increased conversion loss when compared to reactive multipliers. Since most reactive multipliers are narrow-bandwidth devices and require filtered bias circuits [1], the time delay through them is longer than with the resistive types.

Switched multipliers. In Fig. 3-4, a group of multipliers, filters, and switches are configured to increase the frequency of a source f_s, with minimum tuning step Δf_s, to higher values at $2f_s$, $4f_s$, and $8f_s$. As a consequence of using this approach, our minimum tuning step has also been increased from Δf_s to $2\Delta f_s$, $4\Delta f_s$, and $8\Delta f_s$. The circuit path from the output at $2f_s$ to the source f_s has the least number of time delays ($\times 2$, filter and switch), allowing it to have the fastest tuning time whenever the frequency of source f_s is changed.

For the same reason that dividers improve spectral purity by lowering modulation index, frequency multipliers degrade spectral purity by increasing the modulation index by a factor of N, where N is the multipli-

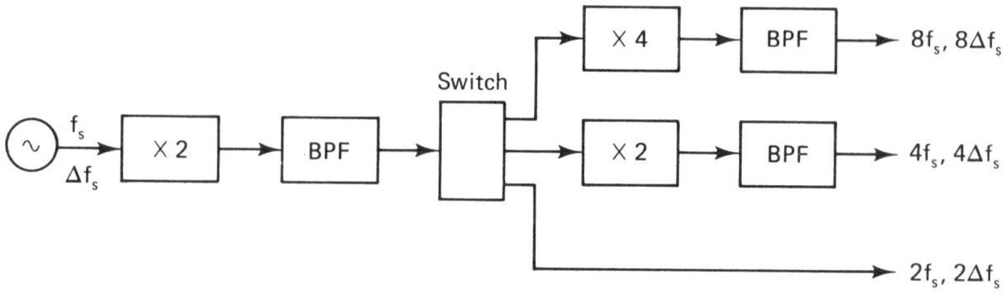

Figure 3-4. Switched frequency multipliers

cation value. At the multiplied output, the carrier-to-noise ratio expressed in dB is reduced by 20 log N. For this reason, highest spectral purity is obtained when low values of frequency multiplication are used in the synthesizer design.

Synthesis with impulse generators. Frequency multipliers derived from step-recovery diodes can be designed to have an output containing harmonics greater than 100 at useful power levels when operating as an impulse generator. Since the wanted group of harmonics could be quite large, a group of switches and filters would not be desirable in light of cost and complexity considerations. Selection of the desired harmonic in microwave synthesizers is usually accomplished by connecting the impulse generator's output to an electronically tunable filter such as a varactor- or YIG-tuned device. When tuning speed to select a particular harmonic is the primary consideration, varactor-tuned types are used since these filters can be tuned at a much faster rate than YIG types.

One potential problem exists when using these types of electronically tuned filters: Tunable filters can degrade spectral purity if spurious voltages or currents are present in the variable control voltage or current applied to the filter. These unwanted responses can cause the filter to phase-modulate the filter center frequency a small amount at rates and deviations set by the control voltage spurious energy levels. At the center of the filter, we can use filter group delay to estimate peak phase deviation:

1. Let K_f in MHz/V be the filter tuning sensitivity.
2. Let τ_d represent filter group delay in radians/MHz. Peak phase deviation $\Delta\theta_p$ is calculated for spurious V_p:

$$\Delta\theta_p = K_f \tau_d \Delta V_p \tag{3-1}$$

Equation (3-1) can be expanded to include the conditions where ΔV_p is a low-frequency ripple or the equivalent noise voltage in a given bandwidth.

Frequency converters. So far we have reviewed the effects on spectral purity and minimum frequency tuning increment when a band-limited signal is raised or lowered in frequency by using multipliers and dividers at the signal source output. Translation to other frequency bands can also be accomplished with mixers without affecting the minimum tuning increment; however, spectral purity can be severely degraded if specific guidelines for mixers are not followed. Before we examine translation with mixers, we need to review the relationships that can cause mixer-generated spurious signals to fall within our desired synthesized passband [2]. In the following discussion, we assume that double-balanced mixers are used and LO signal power (f_l) is larger than input signal power (f_r) by more than 20 dB in order to minimize in-band spurious responses. With these two assumptions as a starting point, we focus our attention on the three most important classes of frequency translators using the following general equation applied to mixers:

$$f_o = nf_l \pm mf_r \tag{3-2}$$

Case 1: Downconversion

$$f_o = f_l - f_r \tag{3-3}$$

with

$$f_l > f_o \quad \text{and} \quad f_r > f_o$$

When f_l and f_r are greater than f_o, no individual harmonics of either input (f_l or f_r) can be in the desired output band; only harmonic differences in the following equation can cause an interference f_i:

$$f_i = nf_l - mf_r \tag{3-4}$$

To minimize the number of potential f_i's, it is necessary to make the ratio of f_r/f_l greater than approximately 0.85, as shown in the derivation of Case 1 in the Appendix.

Case 2: Downconversion

$$f_o = f_l - f_r \tag{3-3}$$

with

$$f_l > f_o \quad \text{and} \quad f_o > f_r$$

For these conditions, we must avoid having f_o near a harmonic of f_r. When this condition cannot be maintained, then we want $f_o >> f_r$ and the ratio given by f_o/f_r to be an even integer. By meeting these conditions, we can make the power level of f_o more than 100 dB above that of f_i. The main reasons are:

1. In a double-balance mixer, even harmonics of the inputs are suppressed much greater than odd harmonics.
2. Even harmonics of the lower power level input signal f_r decrease very rapidly at the higher harmonic levels.

To reduce the f_i's present in Eq. (3-4), $f_i = nf_l - mf_r$, it is necessary to make the ratio of f_o/f_l greater than approximately 0.85 as shown in Case 2 in the Appendix.

Case 3: Upconversion

$$f_o = f_l + f_r \tag{3-5}$$

with

$$f_o > f_l \quad \text{and} \quad f_o >> f_r$$

Under these conditions, the output signal frequency will be slightly larger than the LO frequency at f_l; however, this is a highly desired condition because no harmonics of f_l can be in-band or near the passband. It is still possible for f_o to be equal to a harmonic of f_r; however, our previous discussion for Case 2 (make f_o/f_r an even integer) also applies to this potential condition of Case 3.

To reduce the f_i's present in Eq. (3-2), the general equation, it is necessary to make the ratio of f_l/f_o greater than approximately 0.88 as shown in Case 3 in the Appendix. As this ratio is made larger, f_i's caused by harmonic mixing are reduced; however, as the ratio approaches unity, feedthrough of the LO (f_l) into the output port is present, and as f_o approaches the value of f_l, output filtering of the LO feedthrough becomes more difficult.

While in the process of formulating a particular synthesis approach, it is useful to use one of the many available mixer charts to determine graphically the potential interfering harmonic modulation products; the charts in the Appendix are useful for this purpose. Once the design is finalized, computer programs are used to calculate all interfering signals for any order of input harmonics.

Spectral considerations. In addition to the problems of mixer-generated, in-band spurious signals, mixers also translate the individual power spectral density of noise associated with each input signal to the

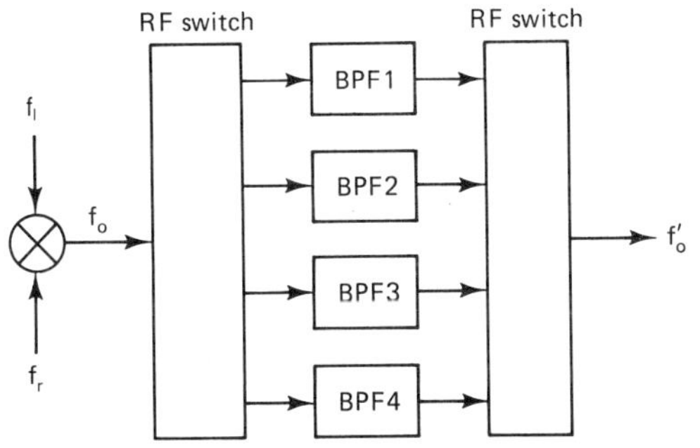

Figure 3-5. Frequency translation with mixer and bandpass filters

output signal. In Fig. 3-5, f_l represents the LO oscillator input and f_r the lower power signal input. The output at f_o contains the sum and difference of f_l and f_r present in the product of (sin $2\pi f_l t$) (sin $2\pi f_r t$) in addition to feedthrough of the inputs to the output. If we consider phase noise modulation of f_l and f_r to be present, the sum and difference frequencies will contain each of the input noise perturbations modulating the output frequency f_o.

If either frequency f_l or f_r is changed rapidly to a new value, the time for this change to appear as f'_o depends on the delays through the mixer, switches, and filters. Since double-balanced mixers can have very wide bandwidths, their delays can usually be neglected, and only the switch and filter delays need to be considered.

COMBINATIONS OF MIXERS, FILTERS, MULTIPLIERS, AND DIVIDERS

In this section, we will show how a simple combination of ten signal sources, RF multipole switches, frequency multipliers, and mixers can be configured to synthesize 1000 discrete frequencies. The block diagram to accomplish this operation is shown in Fig. 3-6. Bandpass filters are required at each multiplier and mixer output; however, they have been deleted from the block diagram in order to concentrate the discussion on synthesis techniques.

The three multipole switches (S1, S2, and S3) are designed to select one of the ten signals present at each input (10.0, 10.1, 10.2, 10.3, 10.4, 10.5, 10.6, 10.7, 10.8, and 10.9 MHz) and apply it to the switch output. We assume that control of these switches is by electronic voltage levels

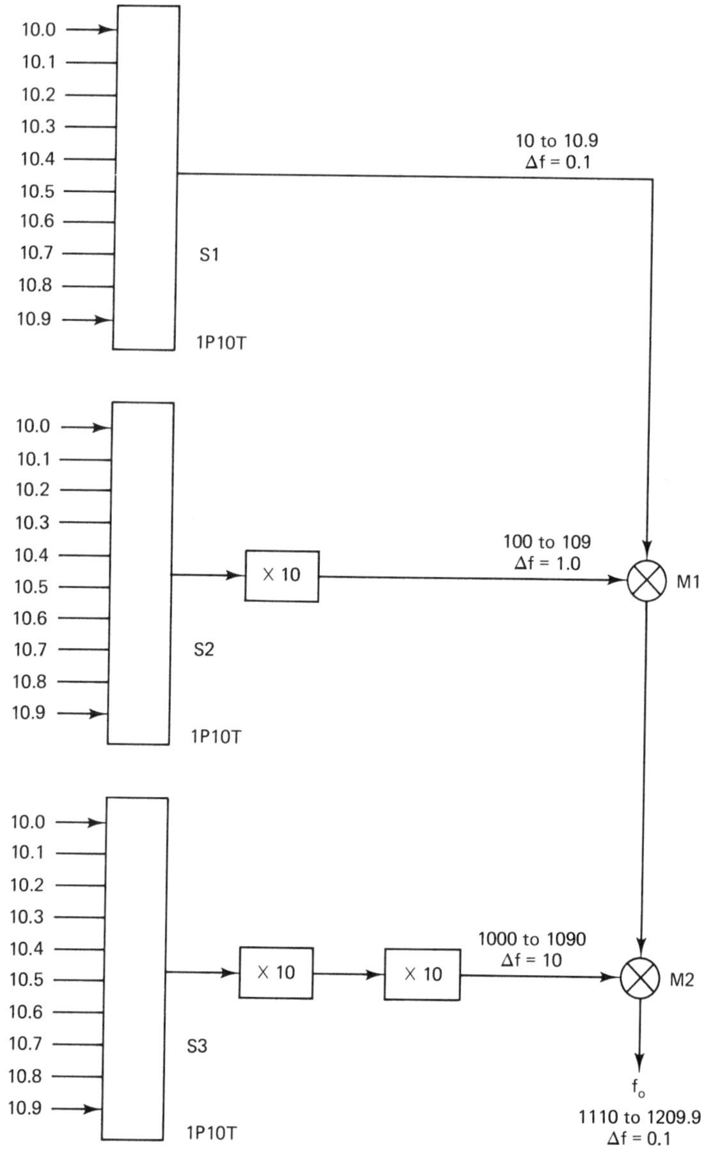

Figure 3-6. Synthesizer with 1000 selectable frequencies. Note: all frequencies in MHz.

applied to command lines to each switch. As a starting point, assume that S1, S2, and S3 are set to select each of the 10-MHz inputs for switch outputs.

At mixer (upconverter), M1, the output frequency is 110 MHz (100 + 10), and this signal is applied to M2 in addition to the 1000-MHz signal derived from the two ×10 multipliers. The final output, f_o, is the sum of (1000 + 100 + 10) MHz, equal to 1110 MHz. Now let us assume that switch S1 is programmed to select the 10.1-MHz input; for this condition, the output at f_o will change from 1110 to 1110.1 MHz. The time required for this frequency step to occur at f_o will be determined by the delay time through S1 and the time delays through the filters at M1 output and M2 output. For this example, it is obvious that S1 controls frequency tuning increments of 0.1 MHz, S2 controls tuning increments of 1 MHz, and S3 controls tuning increments of 10 MHz. A summary equation for the output is given by

$$f_o = (1000 \text{ to } 1090) + (100 \text{ to } 109) + (10 \text{ to } 10.9) \text{ MHz} \qquad (3\text{-}6)$$

with

1000 to 1090	in 10-MHz minimum increments
100 to 109	in 1-MHz minimum increments
10 to 10.9	in 0.1-MHz minimum increments

The time delay required for inputs at S2 and S3 to make changes in the frequency of f_o will be caused by filters on the output of each ×10 and the filter at each mixer output. In Chapter 2 we defined delay time (τ_d) as the 0% to 10% of final amplitude. Since each filter in the signal path has a unique τ_d and τ_r, the total time required for a signal to propagate through several filters will be the sum of each filter's delay τ_d. The risetime (τ_r) is not the sum of each filter τ_r since each filter is not subjected to a step function applied to it. The first filter in each path is the only one that is subjected to the step function. The output of the second filter has a risetime determined by the first and second filters, and so on. In calculating the total risetime, a reasonable approximation is obtained by taking the square root of the sum of the squares of each filter's risetime for a step function input to each filter.

In Fig. 3-7, we have defined the filters to have either 100- or 10-ns risetime, based on required bandwidths of either 10 or 100 MHz. For step frequency changes of 0.1 MHz in the 10- to 10.9-MHz input, the total risetime at f_o is given by

$$\tau_r\,(0.1) = \sqrt{100^2 + 10^2} \text{ ns}$$

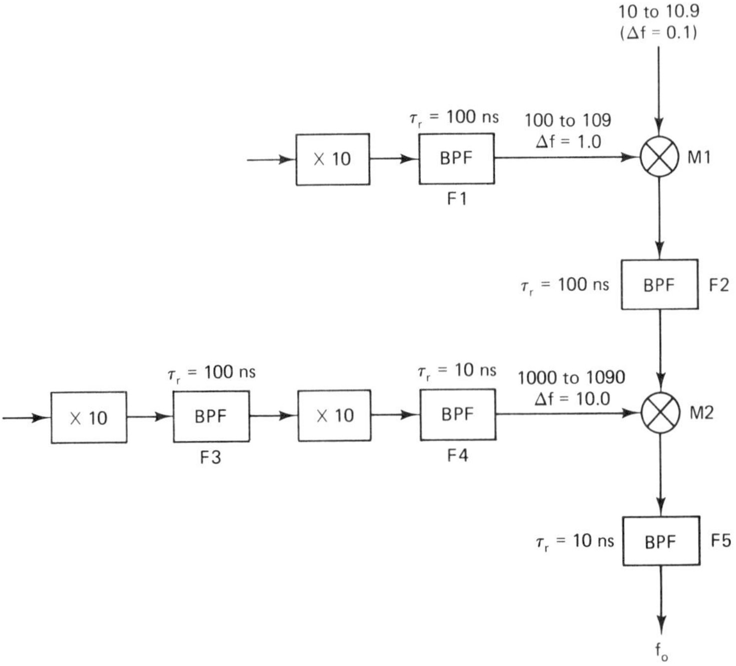

Figure 3-7. Time delays of signal path filters

through F2 and F5. For 1-MHz step change in frequency, the risetime is given by

$$\tau_r\,(1) = \sqrt{100^2 + 100^2 + 10^2}\ \text{ns}$$

through F1, F2, and F5. For 10-MHz step changes in frequency, the risetime is given by

$$\tau_r\,(10) = \sqrt{100^2 + 10^2 + 10^2}\ \text{ns}$$

through F3, F4, and F5.

One additional consideration of this design example is a requirement for spurious sidebands, usually a specified level (in dB) below the desired output signal. The major contributors will be mixer-generated spurious, which have been discussed, and switch isolation of inputs that are not switched to the output.

If a requirement for spurious to be 60 dB down at f_o (Fig. 3-6) were specified, the switch isolation required would be 60 dB for S1. However, S2 and S3 have frequency multipliers in their output path and multipliers increase spurious levels by 20 log N, where N is the multiplier value. For this reason, the isolation required at S2 is 60 dB + 20 log 10 = 80 dB, and the required isolation at S3 is 60 dB + 20 log 100 = 100 dB.

One final comment on mixer spurious outputs that are not in-band: The most important mixer input causing problems in the output is LO feedthrough, and it will usually be close to the desired output band. Typically, we can expect 20 to 30 dB of isolation from the LO port to the output port of the mixer. With an LO at +10 dBm and 25 dB of isolation, the LO feedthrough level is −15 dBm. If we let the low-level input be −10 dBm and assume conversion loss at 7 dBm, the desired output is −17 dBm and is 2 dB lower than the LO feedthrough. In order to have this spurious signal more than 60 dB down, a bandpass filter with many sections is required, and this causes the delay time τ_d to increase as more sections are added to the filter. The filter selectivity can be relaxed if there is a greater frequency separation between the two inputs; however, this would cause more mixer-generated spurious outputs in the passband since the ratio of f_l/f_o decreases as input frequency separation is increased. For these reasons, a compromise between minimizing LO feedthrough and harmonically generated mixer spurious outputs has to be considered when designing frequency converters for direct-type synthesizers.

REFERENCES

1. Penfield, P., and Rafuse, R. P., *Varactor Applications*. Cambridge, Mass.: MIT Press, 1962.
2. Markel, J. D., "Shrinking Intermodulation," *Electronic Design News* (August 1967).

4

Arithmetic Iteration Techniques

Direct-type synthesis can be accomplished with a wide variety of techniques; however, when hundreds of thousands or millions of selectable discrete frequencies are needed, the more efficient iterative circuit approach is employed. This synthesis technique minimizes the number of different circuits and frequency sources required by the synthesizer. The more common variations of this technique use the base 10, 8, 4, or 2 number system. In each of the examples to follow, the amplifier and bandpass filters required at each mixer output are not shown; neither are the filters that are normally part of each frequency multiplier or divider output circuit. We do this to keep the block diagrams uncluttered and relevant to the technical approach illustrated in Fig. 4-1.

BASE 10

In Fig. 4-1, each mixer has one input from a 1P10T RF switch output where one of the ten input frequencies to the switch is selected and present at the switch output. As shown at the top of this diagram, a switch (S_i) can

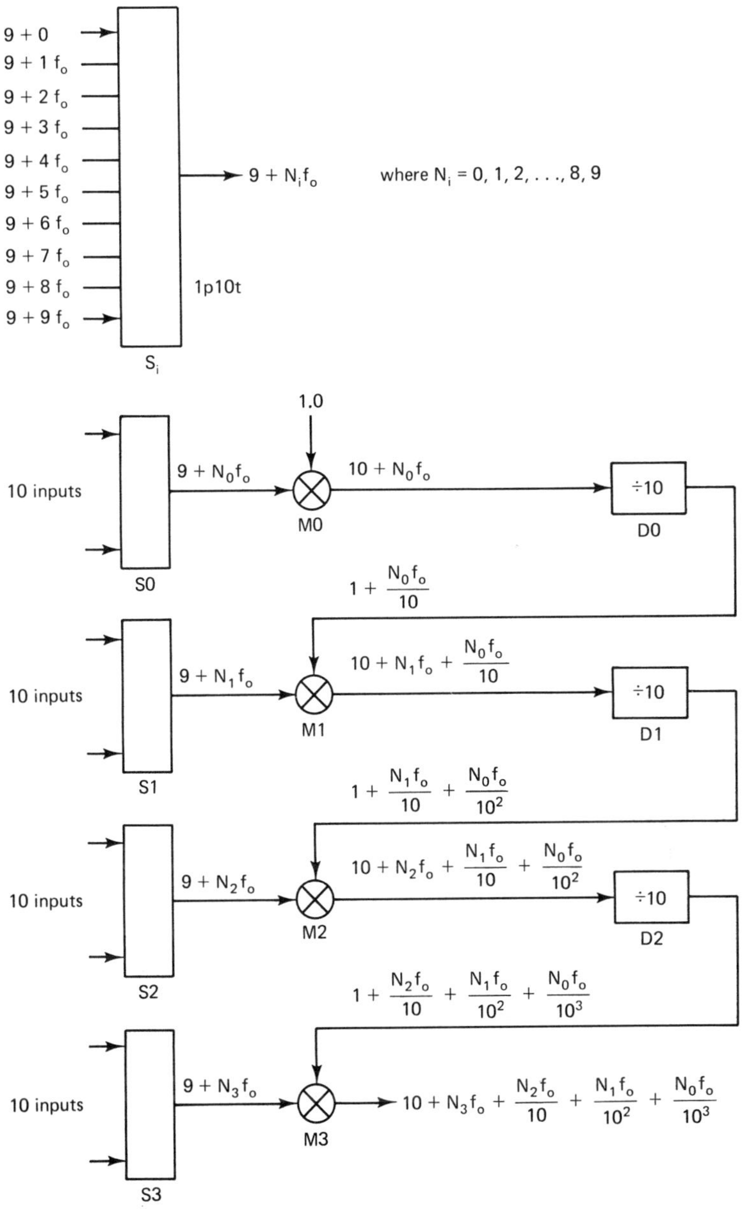

Figure 4-1. Base-10 iteration synthesis technique

have only one output at a time represented by the normalized value 9 + $N_i f_o$, where N_i represents a multiple of f_o, and f_o is equal to 0.1 in this example.

Starting with mixer M0, assume that switch S0 is programmed to select the input line where $N_0 = 0$. For this condition, the output of M0 is $9 + 1 = 10$, and after division by 10, the input to M1 from divider D0 is unity. If we make all remaining switches select the $N_i = 0$ input line, the final output at M3 will simply be 10.

Now let switch S0 select the $N_0 = 1$ line; the output at M0 will change from $10 + 0 f_o$ to $10 + f_o$. At the divider D0 output, the frequency is now $1 + f_o/10$. If we make S1, S2, and S3 select the $N = 0$ line, the only change in the output at M3 is $f_o/10^3$, because the input frequency steps, from S0 output to M0 input, are divided by 10 three times before being applied to M3; therefore, the tuning increment in the final output is determined by the number of dividers present in the path from a switch output to the final output of M3.

At this time, it is informative to compare our previous synthesizer in Fig. 3-6 to Fig. 4-1 by changing the normalized approach to a 1000-MHz output with 100 MHz of bandwidth; to do this, multiply all numbers in Fig. 4-1 by 100 MHz as follows:

$$\text{Output frequency} = 100\ \text{MHz}\left(10 + N_3 f_o + N_2 \frac{f_o}{10} + N_1 \frac{f_o}{10^2} + N_0 \frac{f_o}{10^3}\right)$$

Next let $f_o = 0.1$:

$$\text{Output frequency} = 1000\ \text{MHz} + N_3\, 10\ \text{MHz} + N_2\, 1\ \text{MHz} + N_1\, 0.1\ \text{MHz} + N_0\, 0.01\ \text{MHz}$$

We now have a synthesizer with 99.99 MHz of bandwidth with 10,000 selectable frequencies in 0.01-MHz tuning increments. To acquire this capability, it took 11 reference frequencies, 4 switches, 4 mixers, and 3 dividers. In our example of Fig. 3-6, we needed 10 reference frequencies, 3 switches, 2 mixers, and 3 multipliers.

The major reason for using the iterative technique is that the frequency tuning increment can be decreased without limit by adding the common circuit consisting of a switch, a mixer, and a divider. With the addition of each of these circuit functions, the tuning increment at the output is reduced by a factor of 10. Another significant advantage of this technique is that a high degree of identical circuits are used in the design; that is, all switches are identical, all mixers are identical, and all dividers are identical. The number of tunable discrete frequencies is equal to 10^S, where S is the number of 1P10T switches.

This base 10 synthesis technique has been widely used in synthesizer design because of its simplicity [1]. Typical times required to change frequency ranged from a few microseconds (for large increments) to 30 μs (for very small increments); however, the upper limit of these dividers is limited to lower output frequencies than other base systems.

The time required for a change in output frequency to occur after a switch has selected a new input frequency depends on the total delays encountered as the signal moves to the output. As shown in Fig. 4-2, each mixer output contains three major sources of delay:

1. Bandpass filter delay τ_b
2. Divider delay τ_d
3. Low-pass filter delay τ_l

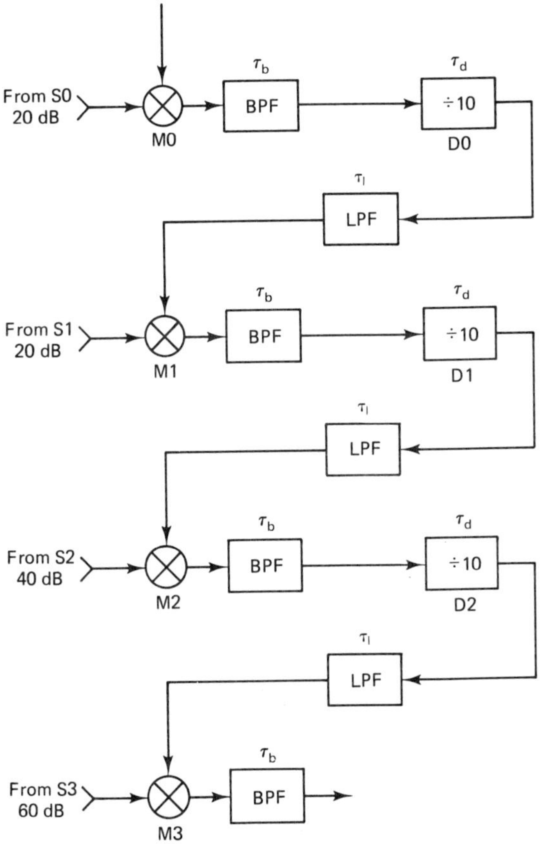

Figure 4-2. Time delays associated with base 10 synthesis

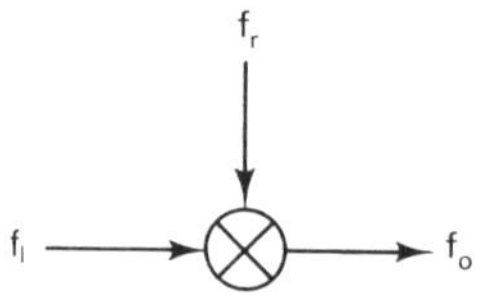

Base 10	Base 8
$f_l = 9$	$f_l = 7$
$f_r = 1$	$f_r = 1$
$f_o = 10$	$f_o = 8$

Figure 4-3. Base-10 and base-8 similarities

If we assume that the mixers are broadband devices (essentially no delay), and switch delays are small compared to the sum of $\tau_b + \tau_d + \tau_l$, the delay for tuning increments is as follows:

1. Multiples of 10 MHz (S3) $= \tau_b$
2. Multiples of 1 MHz (S2) $= (\tau_b + \tau_d + \tau_l) + \tau_b$
3. Multiples of 0.1 MHz (S1) $= 2\,(\tau_b + \tau_d + \tau_l) + \tau_b$
4. Multiples of 0.01 MHz (S0) $= 3\,(\tau_b + \tau_d + \tau_l) + \tau_b$

From these relationships, we note that the delay time increases as we add iterative circuits to obtain smaller tuning increments.

To achieve a given level of spurious rejection at the output (due to switch isolation), the divide-by-10 circuits offer the advantage of decreasing input spurious by 20 dB (20 log 10). This allows the switches farthest from the output to have the least amount of isolation requirements, as noted on the left side of Fig. 4-2. We do not allow the isolation to become too low or the dividers would not function properly. Based on the spurious reduction obtained from the dividers, we could also reduce the complexity of filters that have more than one divider in the path to the output.

Before we proceed to a base 8 synthesizer technique, we need briefly to review our previous discussion on frequency converters (Chapter 3, Case 3). Figure 4-3 shows that for base 10 and base 8, $f_o > f_l$ and $f_o >> f_r$. These conditions reduce the potential number of in-band or near-band spurious responses from mixer-generated harmonic mixing, as discussed previously, when the ratio of f_l to f_o approaches unity.

BASE 8

The base 8 synthesis technique is very similar to base 10, as compared in Fig. 4-4. The main advantages of base 8 synthesis are these:

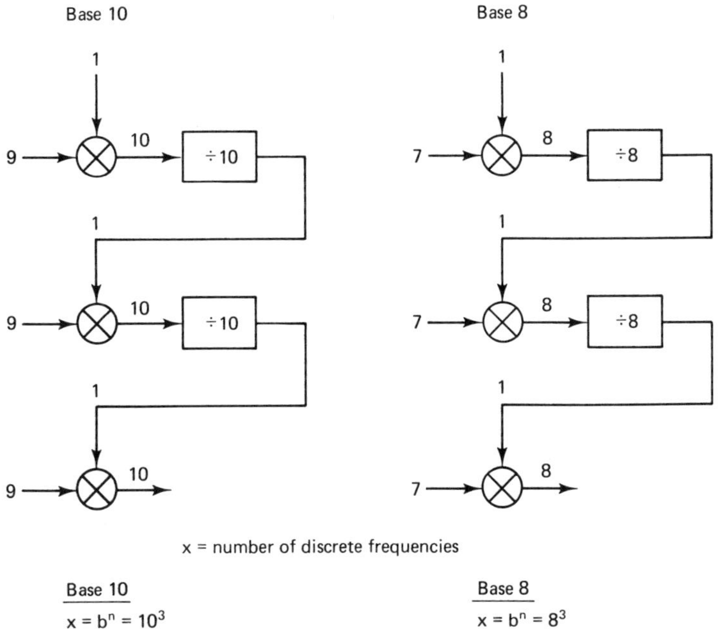

Figure 4-4. Base-8 compared to base-10 synthesis

1. Division by 8 can be three stages of divide-by-2; divide-by-2 has the highest input operating frequency of all other dividers.
2. RF switch 1P8T is easier to implement (combination of two-pole and four-pole) than 1P10T at microwave frequencies.
3. Nine reference frequencies versus 11 for base 10.

To convert our previous base 10 example (Fig. 4-1) to a base 8 example, we need only make a few changes, as done in Fig. 4-5:

1. Replace divide-by-10 with divide-by-8.
2. Replace 1P10T with 1P8T.
3. Reduce switch input frequencies from 10 to 8.

All of our previous discussions on switching time and spurious outputs for base 10 are applicable to base 8 since these two techniques are nearly identical.

We can convert the normalized approach of Fig. 4-5 to any reasonable output band as shown in the following examples:

1. Let the lowest frequency output be equal to 1024 MHz.
2. Let the largest tuning increment be in multiples of 8 MHz.

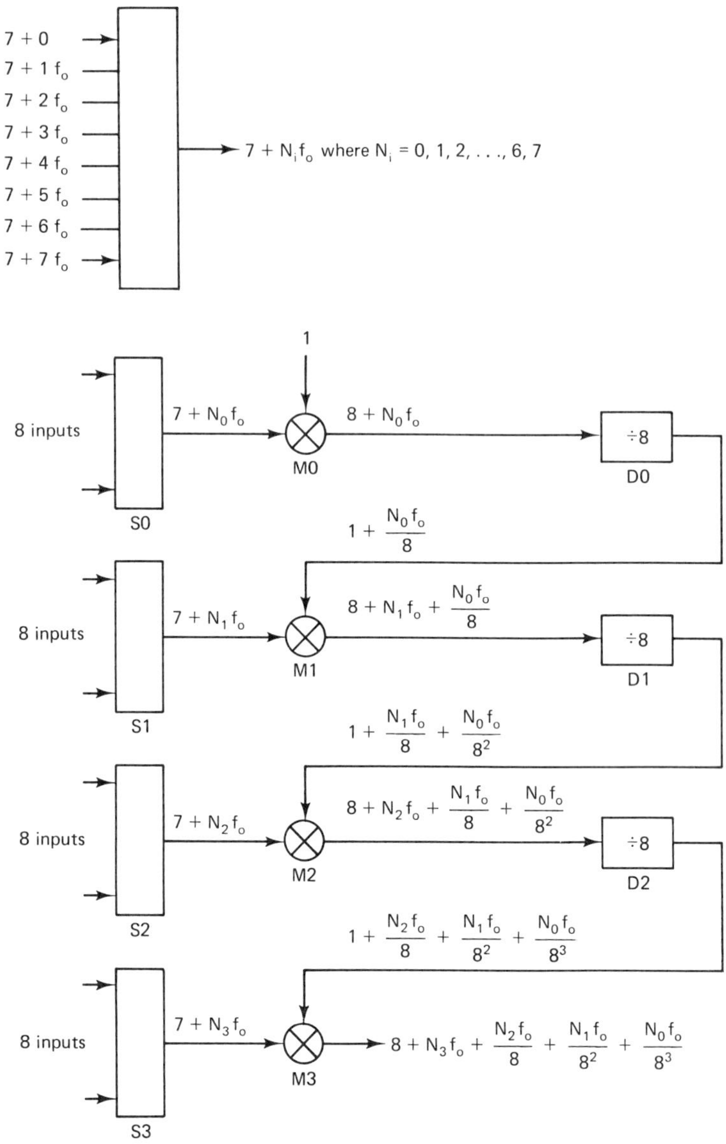

Figure 4-5. Base-8 iteration synthesis technique

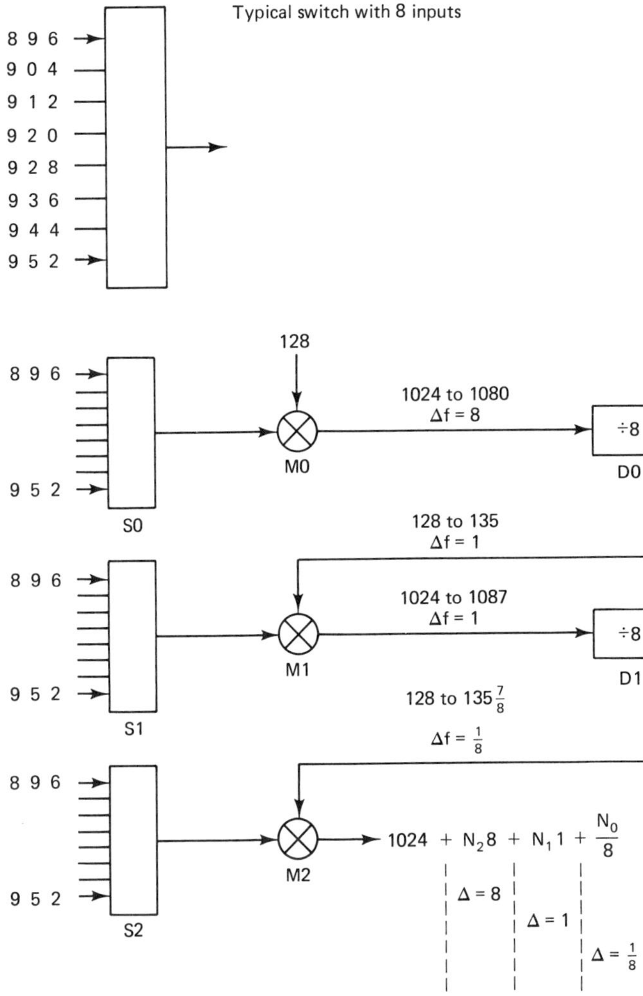

Figure 4-6. Example of base-8 synthesis technique

In Fig. 4-6, for the largest tuning increment to be 8 MHz, f_o of Fig. 4-5 is equal to 8 MHz. For the output to start at 1024 MHz, the input is given by 1024/8 MHz = 128 MHz. The switch inputs are given by:

1. 7 × 128 = 896 MHz
2. 896 + 8 = 904
3. 896 + 16 = 912
4. 896 + 24 = 920

5. 896 + 32 = 928
6. 896 + 40 = 936
7. 896 + 48 = 944
8. 896 + 56 = 952

From inspection of Fig. 4-6,

1. The number of discrete frequencies is given by 8^3.
2. The minimum tuning increment is equal to 8 MHz/8^2.
3. The output is tunable from 1024 to $1087\frac{7}{8}$ MHz.

Comments on spurious outputs and the time delay resulting from tuning to different frequencies is essentially the same as with regard to Fig. 4-2. In that discussion, we determined that as more "mix and divide" stages are added to increase tuning resolution, the delay time of these smaller frequency steps is increased. In addition, as more divide-by-8 stages are in the signal path to the output, the requirements for switch isolation can be relaxed by 20 log N, where N is the total amount of division in the signal path to the output.

BASE 2

Synthesizer designs using a base 2 technique offer the greatest potential for minimum switching time to change the output from one frequency to another. The major reason is that divide-by-2 circuits can be designed to operate at inputs greater than 4 GHz; for this reason, larger bandwidths are potentially available at the higher frequency of operation. In Fig. 4-7, a normalized model of the common circuit is illustrated.

The normalized fixed inputs are designated 10 (on the left side), 2 (on the top), and 9 (on the right side), with

1. ΔL an incremental increase in left-side input (10)
2. ΔT an incremental increase in top input (2)
3. ΔR an incremental decrease in right-side input (9)

For the conditions shown in Fig. 4-7, the output is related to the inputs as noted at the bottom of the figure. We can summarize the output-input relations as follows:

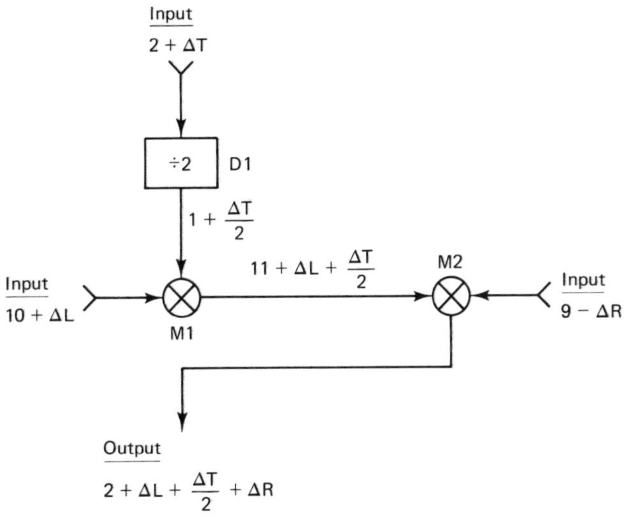

Figure 4-7. Base-2 synthesizer circuit

1. Increasing the inputs to mixer M1 causes the output to increase (the same positive sense).
2. Decreasing the right-side input to M2 causes the output to increase (inverted sense).

Before proceeding to examples of base 2 synthesis, we need to examine the reason for having two mixers per frequency divider instead of one per divider as in base 8 or base 10. The reason for doing this is to prevent the mixer's output from containing excessive spurious outputs. Two possible configurations for a common circuit are illustrated in Fig. 4-8. The main problem with either of these frequency plans is that the input values will cause many mixer-generated spurious responses to be in or near the desired output band. This is why the base 2 (binary) synthesizer circuit

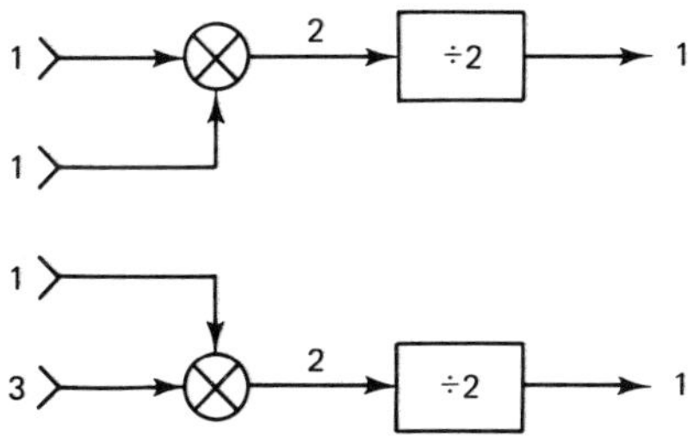

Figure 4-8. Mixers with excessive spurious outputs

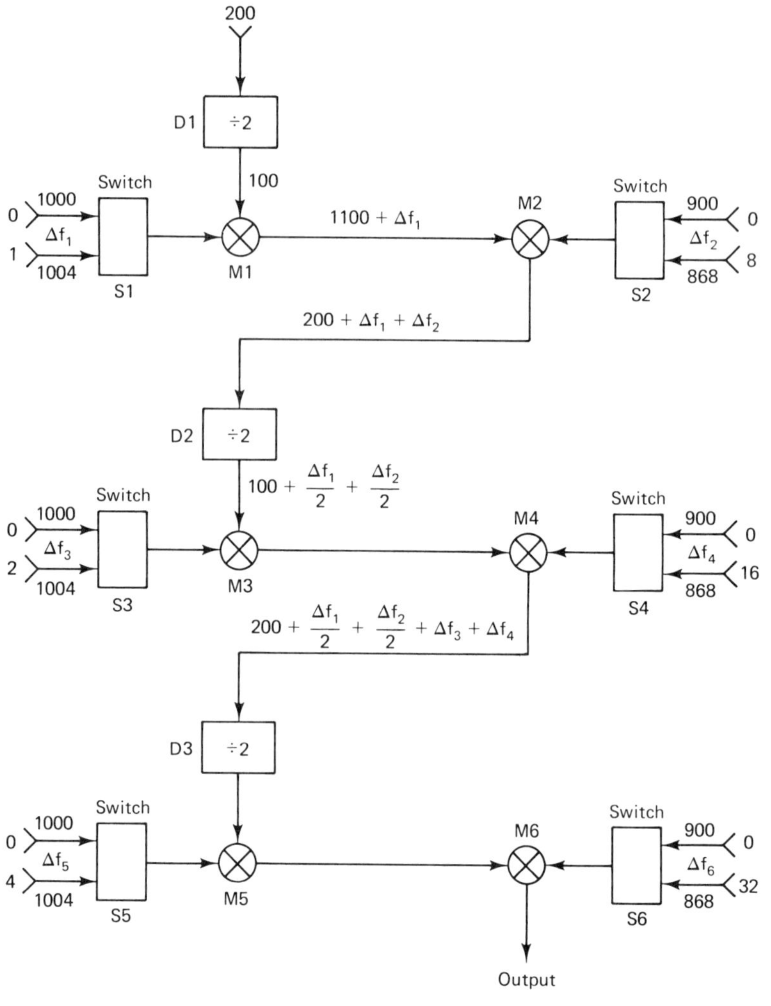

$$\text{Output} = 200 + \Delta f_5 + \Delta f_6 + \frac{\Delta f_3 + \Delta f_4}{2} + \frac{\Delta f_1 + \Delta f_2}{2^2}$$

Figure 4-9. One version of base-2 synthesizer

requires an up-down translator, as shown in Fig. 4-7. Incidentally, the inputs designated 10 and 9 in Fig. 4-7 could be slightly lower or much greater than the selected values; these values are selected to emphasize the following:

TABLE 4-1

Switch	Output	Value
S1	$\Delta f_1 \div 4$	1
S3	$\Delta f_3 \div 2$	2
S5	Δf_5	4
S2	$\Delta f_2 \div 4$	8
S4	$\Delta f_4 \div 2$	16
S6	Δf_6	32

1. At M1 (an upconverter), the LO input frequency must be much greater than the signal input frequency.
2. At M2 (a downconverter), the LO and signal input frequency must be much greater than the output frequency.

The base 2 synthesis technique can have many forms, one of which is illustrated in Fig. 4-9. Input lines to switches are designated as a 0 or a 1 as in the binary number system. With all switches in the 0 path, the 200-MHz input at the top of the figure is divided by 2 to become 100 MHz. It is summed with 1000 MHz to yield 1100 MHz, mixed with 900 MHz, and downconverted to 200 MHz. This process is repeated at each horizontal pair of mixers. Now, let S1 select the 1 path (1004), which corresponds to an incremental increase of 4 MHz (Δf_1) from S1 to M1. Before this increment reaches the output, it is divided twice by 2 to become a 1-MHz increase in the output. Examination of the remaining switch paths for the output at mixer M6 yields the relationship shown in Table 4-1.

Aside from having the potential for fastest switching speed, the base 2 synthesizer has several advantages:

1. Switches are 1P2T and are less complicated than 1P10T.
2. Base 2 has the least number of fixed frequency references (five).
3. Each divide-by-2 reduces spurious by 6 dB.
4. Since divide-by-2 GaAs digital circuits operate at frequencies greater than 4 GHz, base 2 has the highest output frequency capability.

The base 2 synthesizer has one major disadvantage: Since mixers have conversion loss, an amplifier is required for each pair of mixers, and the number of discrete frequencies is given by 2^N (for N mixers); thus it has the largest number of parts (and costs) to achieve its potential for fractional nanosecond switching times between output frequencies.

BASE 4

The base 4 normalized design is very similar to base 2 (compare Fig. 4-10 to Fig. 4-7). The two mixers form an up-down translator with an offset to account for division by 4 within the translator. Like the base 2, inputs to M1 are summed, and increases in either input are upconverted at the output of M1. At mixer M2 output, the downconverted signal will increase when M1 output increases and when the right-side input to M2 decreases.

The main advantage of base 4 synthesis over base 2 synthesis is that the number of selectable frequencies per mixer is twice that of base 2; therefore, for the same number of selectable frequencies, the base 4 synthesizer would have half the number of parts as a base 2 design. This reduction in parts, including filters, may make it possible for the base 4 design to approach the base 2 design in frequency switching speed.

An example of base 4 design is illustrated in Fig. 4-11. Input lines to 1P4T switches are defined in terms of base 4:

S1:	0,	1,	2,	3
S3:	0,	4,	8,	12
S2:	0,	16,	32,	48
S4:	0,	64,	128,	192

As a starting point, let each input to a switch be in the 0 position; this will cause the output of M4 to be 400 MHz. If the input at S1 is changed from

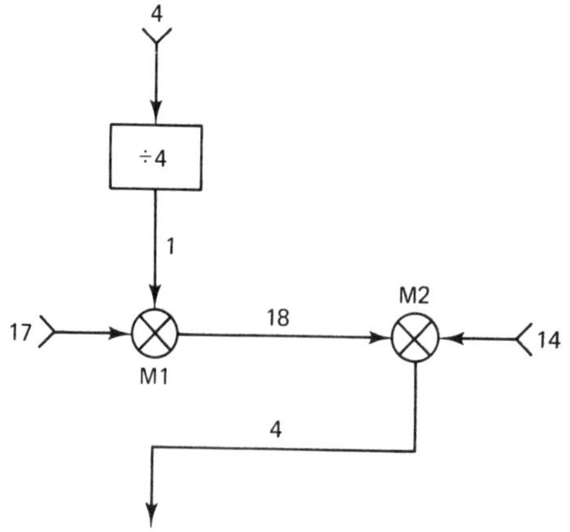

Figure 4-10. Normalized base-4 circuit

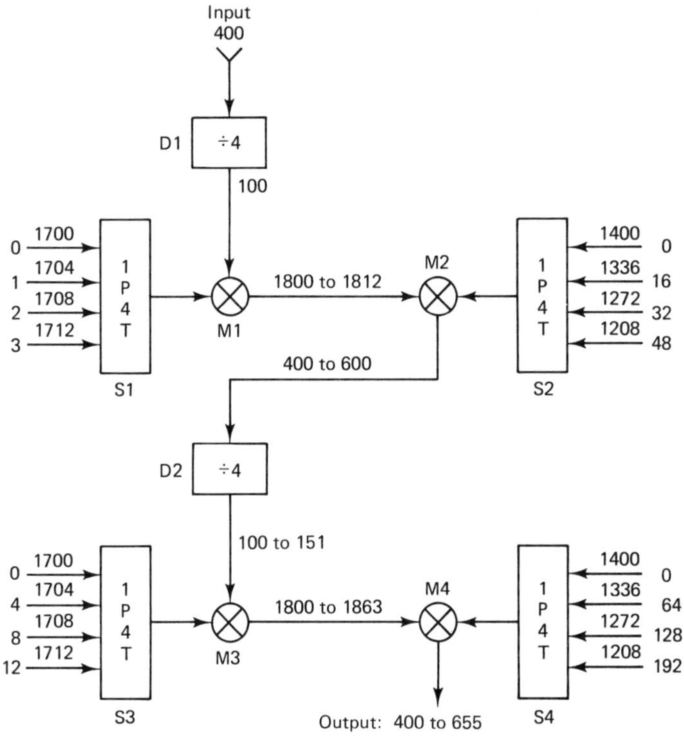

Figure 4-11. Base-4 synthesizer example

1700 to 1704 MHz, the output from M4 will increase by 1 MHz since the 4-MHz increase to M1 is divided by 4 one time in the signal path to M4 output. For this synthesizer, the number of discrete frequency outputs is given by 4^N, where N is the number of mixers or switches in the synthesizer. With $N = 4$, this example has 256 selectable frequencies at the output. If we were to expand the number of mixers from 4 to 16, 8 up-down translators, the number of selectable frequencies would increase to $16^4 =$ 16,536. As with our previous examples of iteration techniques for base 10, 8, and 2, the base 4 switching time is limited mainly by filter bandwidths and the number of filters in the circuit path from each switch input to final mixer output.

Spectral purity is enhanced by 20 log 4 at each divider and improves carrier-to-noise or spurious ratio by 12 dB per divider; therefore, switches with the most number of divide-by-4 circuits in the path to the output can have the least isolation performance.

SUMMARY

The optimum choice of direct-type synthesizer approach using base 2, 4, 8, or 10 that we have considered will depend on the specific application to determine which technique will have the fastest frequency set-on time with lowest associated product cost. The fundamental relationship that determines parts' complexity is given by $X = B^N$, where

X = number of selectable frequencies

B = number base

N = number of switch-mixer combinations

We also need to remember that each mixer has conversion loss and requires filtering to lower spurious responses. If we allow one stage of amplifier gain to make up for the loss in each mixer and filter combination, then we can let N represent the number of iterative stages in the synthesizer.

Let us assume that a synthesizer is required with at least 1 million discrete frequencies. For the techniques we have examined:

Base 10: 10^6 = 1,000,000; N = 6
Base 8: 8^7 = 2,097,152; N = 7
Base 4: 4^{10} = 1,048,576; N = 10
Base 2: 2^{20} = 1,048,576; N = 20

We immediately realize that base 4 has half the number of stages (10) as base 2 (20) and is a better choice; however, base 8 and base 10 have 70% and 60%, respectively, the number of stages as base 4 and should be an even better choice. To make the final decision, we would have to include the number of independent sources with their associated complexity and the relative importance of frequency set-on time to change the output signals' frequency. The number of independent sources in each base is listed in Table 4-2.

Since base 4 and base 8 have the same requirements, the octal-based synthesizer is a better choice. Now let us return to our problem of selecting the optimum synthesis approach when very fast set-on times are required. For speed considerations, base 10 is the slowest and should always remain so in the future because a divide-by-10 will always contain at least a divide-by-2 and a divide-by-5 in series, and divide-by-5 should never be able to operate as high in frequency as divide-by-2; thus we can delete base 10 on the basis of being the slowest. After all, the only reason even to consider direct frequency synthesis is to obtain very fast tuning times. On the basis

TABLE 4-2

Base	Number of Sources
10	11
8	9
4	9
2	5

TABLE 4-3

Base	N	S	Sum
8	7	9	16
4	10	9	19
2	20	5	25

of source complexity, base 2 has five, and this is nearly half the number of base 8 or base 4, which require nine each. In Table 4-3 we list the number of stages (N) required and sources (S) for the three fastest approaches to obtain approximately 1 million selectable frequencies.

If the number of parts required to make up each source is on the same order of complexity as each iterative stage N, synthesizer complexity is related to the sum of $S + N$, as shown in Table 4-3; thus the base 8 approach would have the least complexity.

If the delay time through each iterative stage is nearly the same whether the stage is for base 2, 4, or 8, the base 8 approach has the smallest delay time associated with it. Again we find base 8 a better choice.

One final consideration: As the requirement for discrete selectable frequencies X decreases, the number of stages N is lowered and causes the sum of $N + S$ for all three approaches to become nearly the same. When this condition is reached, the base 2 approach may be the best, since the 1P2T RF switches are simpler in complexity than the higher-order switches of base 4 and 8.

REFERENCE

1. Hewlett Packard Application Note 96.

5

Design Examples with Measured Performance

In this chapter are described two design examples, which were designed, assembled, and tested to verify the design techniques. The measured data and the block diagrams presented in this chapter are similar to the actual hardware developed for use in several applications. The first example was designed using a noniterative approach, as described in Fig. 3-6 and is illustrated for this design example in Fig. 5-1.

EXAMPLE 1

On the left side of Fig. 5-1, the ten RF sources shown are derived from ten independent crystal oscillators, which are not phase-locked to a single common standard. For the application of this synthesizer, having all outputs derived from a single standard was not important. What was needed was high spectral purity with excellent short-term stability in a 1-ms time interval. In addition, the time required to change the output frequency

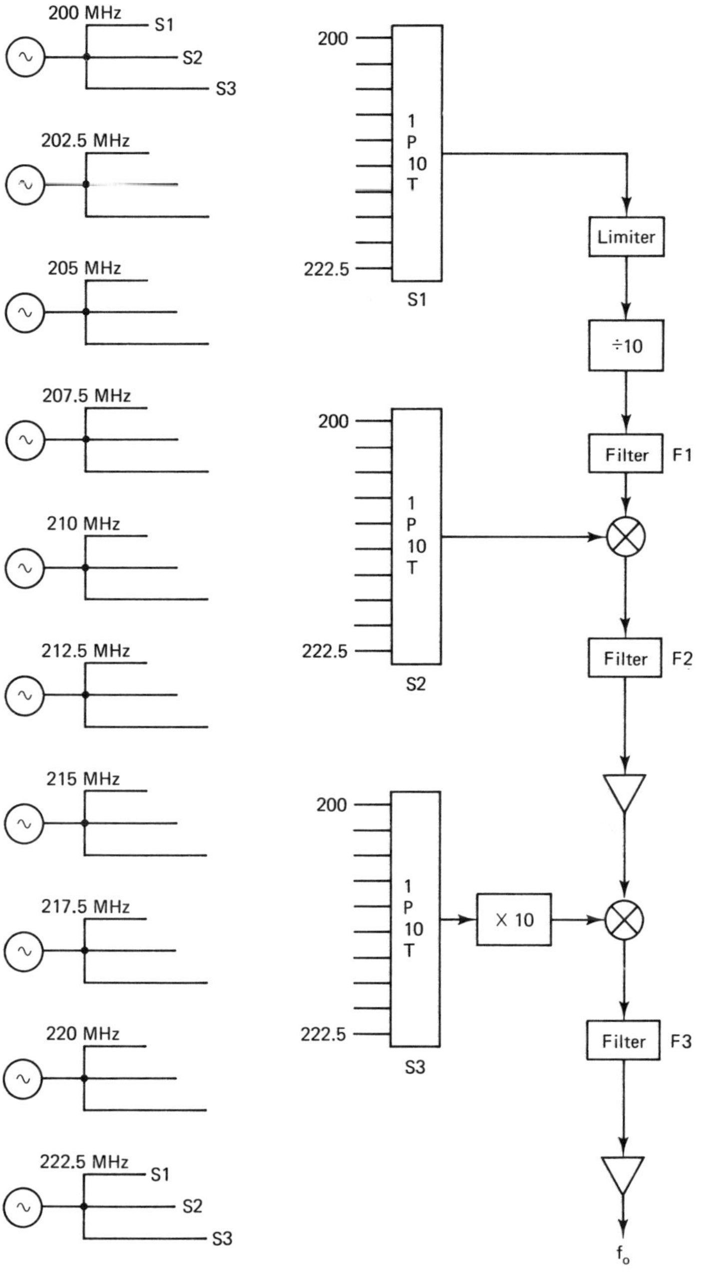

Figure 5-1. Base-10 synthesizer, measured performance

with TTL voltage levels determining the 1P10T switch selection had to be less than 1 μs.

For this synthesizer, the output frequency is determined by

$$f_o = [10 \times (S3)\ \text{output}] + (S2)\ \text{output} + [(S1)\ \text{output}/10] \qquad (5\text{-}1)$$

where S1, S2, and S3 represents the single output frequency selected at each switch (S1, S2, and S3) by TTL logic. The minimum frequency output is

$$f_o\ \text{min} = (10 \times 200) + 200 + (200/10) = 2220\ \text{MHz}$$

and the maximum frequency output is

$$f_o\ \text{max} = 10 \times 222.5 + 222.5 + 222.5/10 = 2469.75$$

with a minimum tuning increment of 0.25 MHz.

Table 5-1 contains a list of the more significant time delays in the longest signal path (input of S1 to f_o). It shows the total time required (delay plus 10% to 90% rise) for making changes in the output frequency from switch S1 inputs. By designing switch S1 for 200 ns maximum to select one of the ten inputs, the "worst case" total time is under 1 μs.

The frequency set-on time was measured in the test fixture diagrammed in Fig. 5-2. In this test fixture, the square-wave generator provides external synchronization of the oscilloscope and determines at which instant the digital controller output is switched between two preset TTL commands for controlling the frequency synthesizer. The two frequencies selected cause switch S1 to cycle between the 200- and 222.5-MHz inputs at S1 and is a tuning increment of 2.25 MHz in f_o. For accurate measurements to be obtained, precise static measurements of the discriminator output voltage are required at the two different input frequencies.

Spectral purity of this design was measured using a microwave discriminator configured from high-Q cavities. The measured spectral purity data are summarized in Table 5-2.

TABLE 5-1

Component	Bandwidth (MHz)	τ_d (delay) (ns)	τ_r (rise) (ns)
÷ 10	N/A	50	N/A
Filter F1	5	200	200
Filter F2	30	33	33
Filter F3	300	3	3
		286	233
Total time = 489 ns			$\sqrt{\tau_r^2}$ = 203

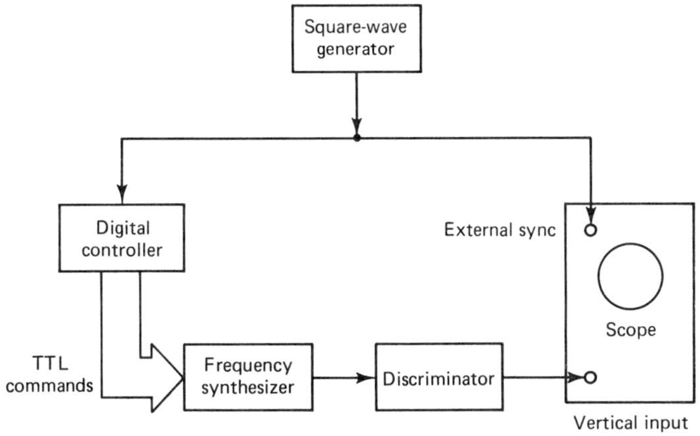

Figure 5-2. Measurement of frequency switching speed with discriminator

EXAMPLE 2

In this example, we examine a technique similar to, but not identical to, a binary-type synthesizer invented by the author (U.S. Patent No. 4,425,552) while employed by the Sperry Corporation, to whom the patent was assigned. A frequency synthesizer was designed and tested by the author on U.S. government funding using a similar technical approach, as shown in Fig. 5-3 [1].

The synthesizer design shown in Fig. 5-3 requires five fixed-frequency sources of 600, 2948.8, 3000, 3450, and 3475.6 MHz, respectively. With switches S1, S3, S5, S7, and S9 controlled to select the 3450-MHz input and S2, S4, S6, S8, and S10 controlled to select the 3000-MHz input, the

TABLE 5-2

Frequency Removed	SSB Noise (dB down/ Hz)	Spurious
100 Hz	80	—
1 kHz	110	—
10 kHz	113	—
100 kHz	120	—
1 MHz	116	—
2.5 MHz	—	88 dBc
5 MHz	—	78 dBc
10.5 MHz	—	82 dBc
20 MHz	—	74 dBc
30 MHz	167	—

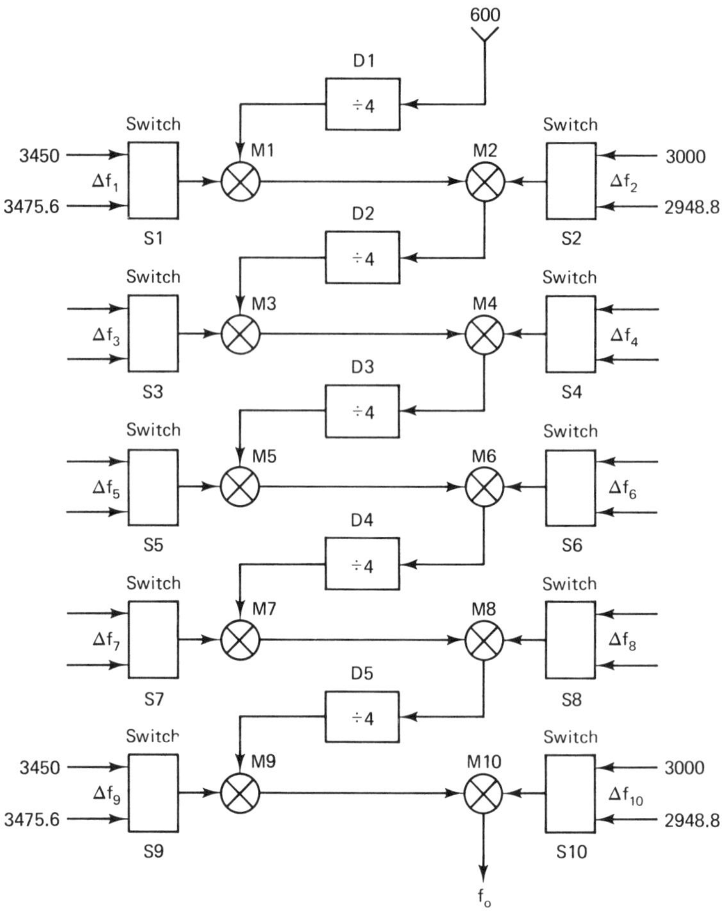

Figure 5-3. Synthesizer, designed and tested

following frequencies are synthesized:

$$\text{M1 output} = 3450 + 150 = 3600 \text{ MHz}$$

$$\text{M2 output} = 3600 - 3000 = 600 \text{ MHz}$$

All mixers on the left side will have the same output as M1, and all on the right side will have the same output as M2. If we now cause S1 to select the 3475.6-MHz input (an increase of 25.6 MHz), the output from M1 will increase from 3600 to 3625.6 MHz. This increase will occur at the output of M2 (3625.6 to 3600 MHz) and increases M2 output to 625.6 MHz; however, the 25.6-MHz increase in frequency is divided by 4 a total of four times in the signal path to mixer M10 and causes the output to increase from 600 to 600.1 MHz.

Although the design features the repeated use of divide-by-4, this synthesizer is a base 2 design since only two selectable frequencies are available at each switch input. If we let Δf_1, Δf_3, Δf_5, Δf_7, and Δf_9 be positive when the selected frequency is changed from 3450 to 3475.6 MHz and let Δf_2, Δf_4, Δf_6, Δf_8, and Δf_{10} be positive when the selected frequency is changed from 3000 to 2948.8 MHz, the output frequency from M10 is equal to

$$f_o = 600 + \Delta f_{10} + \Delta f_9 + \frac{\Delta f_8 + \Delta f_7}{4} + \frac{\Delta f_6 + \Delta f_5}{4^2} + \frac{\Delta f_4 + \Delta f_3}{4^3} + \frac{\Delta f_2 + \Delta f_1}{4^4} \tag{5-2}$$

We can restate this equation in tabular form as in Table 5-3. Thus the M10 output can be set on frequencies starting at 600 MHz and ending at 702.3 MHz in 0.1-MHz tuning increments.

In Fig. 5-4, there are five stages of identical circuits with each circuit containing a divide-by-4, two mixers, and two switches. The common circuit is shown in more detail in Fig. 5-4 with emphasis on those elements that contribute the most to determining the time it takes for a frequency change at the input to occur in the output. The total delay through this circuit from divide-by-4 input to the limiting amplifier output is equal to the sum of each delay time (τ_d) plus

$$\sqrt{\tau_r^2} = \tau_r, \quad \text{total risetime} \tag{5-3}$$

Since a limiting amplifier is in every common circuit, to establish constant voltage levels into each divide-by-4, the delays of each common circuit are added to compute the total synthesizer delay. The total delay per stage

TABLE 5-3

	M10 Output Increment (MHz)
S1 controls Δf_1	0.1
S2 controls Δf_2	0.2
S3 controls Δf_3	0.4
S4 controls Δf_4	0.8
S5 controls Δf_5	1.6
S6 controls Δf_6	3.2
S7 controls Δf_7	6.4
S8 controls Δf_8	12.8
S9 controls Δf_9	25.6
S10 controls Δf_{10}	51.2
Total BW	102.3

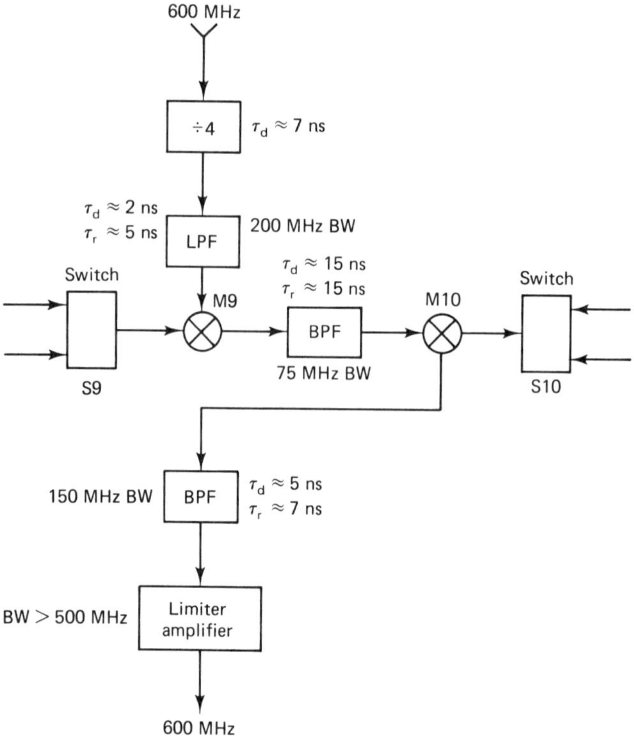

Figure 5-4. Time delay for each identical circuit

is given by the sum of total τ_d + total τ_r:

$$\text{total } \tau_d = (7 + 2 + 15 + 5) \text{ ns} = 29 \text{ ns} \tag{5-4}$$

$$\text{total } \tau_r = \sqrt{5^2 + 15^2 + 7^2} = 17 \text{ ns} \tag{5-5}$$

where

$$\text{delay per stage} = 46 \text{ ns}$$

$$\text{delay of five stages} = 5 \times 46 \text{ ns} = 230 \text{ ns}$$

In order to measure synthesizer frequency set-on time, the test fixture diagrammed in Fig. 5-5 was constructed. In this synthesizer, the five common frequencies of 600, 3000, 2948.8, 3450, and 3475.6 MHz were all phase-locked to a single crystal oscillator at frequency f_s. The switch command lines were configured to make the synthesizer output switch at a 100-kHz rate between 600 and 606.4 MHz. When the synthesizer output is 600 MHz, the output from the double-balanced mixer (DBM) is a DC voltage representing the phase difference between the two 600-MHz inputs.

When the synthesizer output is 606.4 MHz, the DBM output is at the difference value of 6.4 MHz. The picture observed on the oscilloscope was alternating between the DC level and a 6.4-MHz beat frequency.

One problem occurs with this measurement technique: There is an inherent ambiguity of phase coherency caused by the frequency dividers within the common circuit. Since the dividers are not reset to 0 whenever the frequency output is reprogrammed, the dividers can accumulate a count of 0, 1, 2, or 3, and this condition causes the initial phase of each beat frequency output to have as many as 4^N initial phase states where N is the number of divide-by-4's between the switch and synthesizer output. Using a discriminator as outlined in Fig. 5-2 avoids this problem, since it is sensitive to frequency changes and not sensitive to phase changes.

Up to this point in our design example, we have examined a direct-type synthesizer with limited bandwidth (600 to 702.3 MHz). In the last part of this chapter, techniques to increase the bandwidth are described, starting with Fig. 5-6. If we let the output from mixer M10 in Fig. 5-4 be the input to M11 of Fig. 5-6, S11 controls a 102.4-MHz tuning increment and S12 controls a 204.8-MHz tuning increment at the output of mixer 12. As a consequence of this up-down translator stage, the bandwidth is increased from 102.3 MHz to

$$102.3 + 102.4 + 204.8 = 409.5 \text{ MHz} \tag{5-6}$$

with a minimum tuning increment of 0.1 MHz.

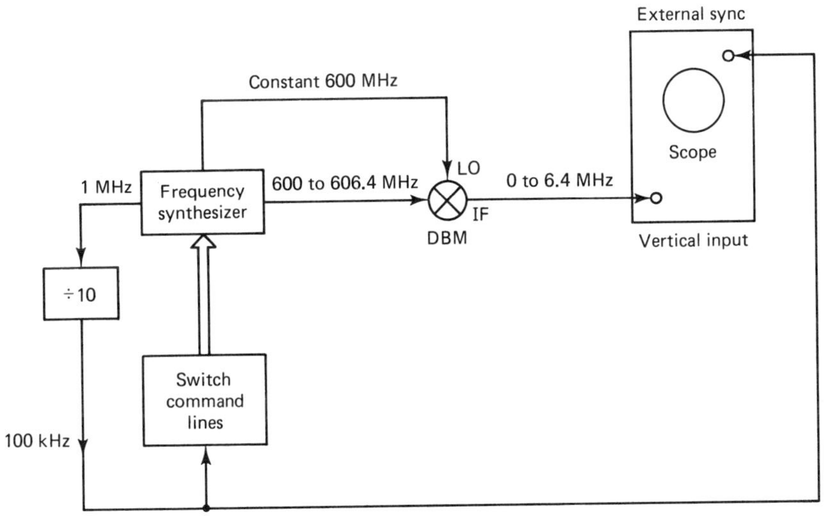

Figure 5-5. Measurement of frequency synthesizer switching speed

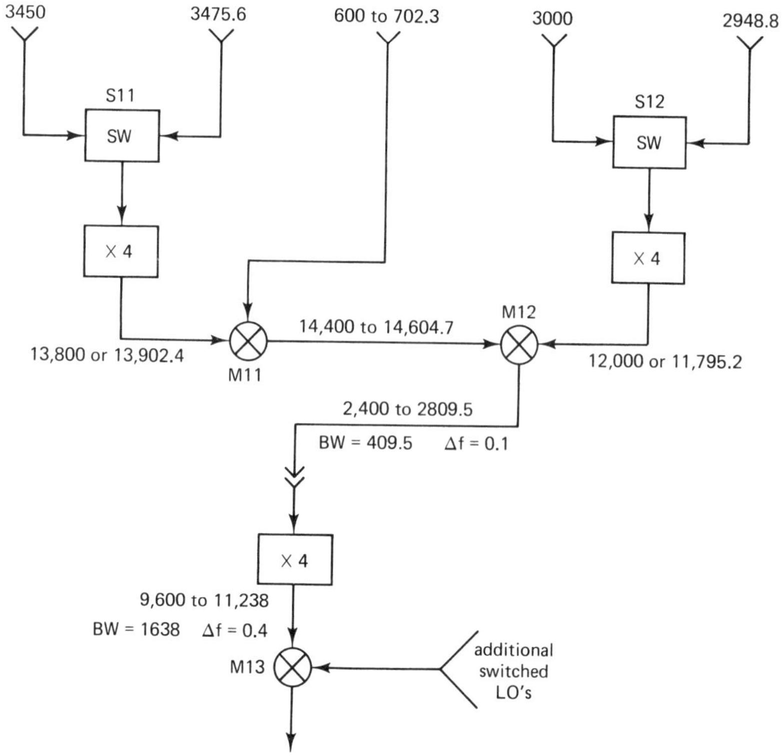

Figure 5-6. Increasing the synthesizer output bandwidth

The bandwidth can still be expanded from a ×4 frequency multiplier (as illustrated in Fig. 5-6) to 4 × (409.5) MHz and is equal to 1638 MHz (11,238 − 9600). Of course, the minimum tuning increment is increased from 0.1 to 0.4 MHz; however, the 0.1-MHz tuning resolution can be maintained by adding another common circuit to the input of divider D1 in Fig. 5-3.

With the additional common circuit, we now have 16,384 selectable frequencies with 0.1 MHz of resolution in the band extending from 9600 to 11,238.3 MHz. The bandwidth of 1638.3 MHz can be translated to lower or higher frequency bands with the addition of more switched LOs that are phase-locked to the synthesizer's single common crystal oscillator. For example, if we have an LO at 9600 MHz, the downconverted output from mixer M13 extends from 0 to 1628.3 MHz. An LO at 8000 MHz would allow the M13 output to be from 1600 to 3238.3 MHz. The number of additional frequency bands that we add to the synthesizer must be selected very carefully to minimize spurious signal generation from the addition of more mixers and frequency multipliers.

REFERENCE

1. Stirling, R. C., Fast Switching Microwave Frequency Synthesizers, *Proceedings of 1979 Southwestern Region Technical Symposium, Association of Old Crows, San Antonio, Texas, 1979*, pp. 73–82.

6

Circuits Used in Indirect-Frequency Synthesis

In this chapter and in Chapters 7 through 13, indirect-frequency synthesizers design techniques are described. Chapter 6 is a brief review of phase detectors and electronically tuned oscillators. In Chapter 7 the operation of an analog phase-locked loop is reviewed using Bode-type graphs to predict stability and spectral purity. Chapter 8 begins with modeling single digital loops and includes the Bode graph with a programmable divider in the feedback path. Chapter 9 describes techniques for designing multiple digital phase-locked loops and the reasons for using multiple loops. In Chapter 10 we examine the more conventional techniques employed in microwave frequency synthesizer designs and finish with an analysis of a design technique invented by this author. Several microwave synthesizer design techniques are presented in Chapter 11 where this design approach is used. In Chapter 12 the technique is expanded to other potential applications and concludes with an additional new approach for microwave synthesizers. Chapter 13 describes measurement of synthesizer performance.

BASIC CIRCUITS OF INDIRECT SYNTHESIS

The circuits employed in direct synthesis designs (frequency mixers, multipliers, and dividers) are sometimes needed for indirect synthesizers when a multioctave bandwidth is required. Internal to the indirect synthesizer with less than an octave bandwidth, the most commonly used devices are programmable dividers (frequency counters) and phase detectors. Phase detectors for analog applications almost always use the double-balanced mixer (DBM) with the applied LO and RF port signal levels at high levels. At these higher power levels, the mixer is driven close to, or in, a saturated mode. With the DBM being driven hard, the applied sinewave inputs are squared and approximate square waves in the internal circuits. Most DBMs will function as phase detectors if the output IF port responds from DC to a high-frequency upper limit. The intent of this author is to discuss a single model of the phase detector, using a saturated DBM, and not concern ourselves with the nonideal characteristics of DC offset, linearity, and sensitivity; these topics are treated adequately in DBM manufacturer's application notes [1].

Our simple model will view the DBM as a product detector that has an IF output, lower in voltage than the inputs, which is directly proportional to the product of the two inputs at a frequency of ω_i and a phase difference of θ_1:

$$\text{IF}(t) = K \times \text{LO}(t) \times \text{RF}(t) \tag{6-1}$$

where

$$K = \text{a constant}$$

$$\text{LO}(t) = \cos \omega_i t - \frac{1}{3} \cos 3\omega_i t + \frac{1}{5} \cos 5\omega_i t \tag{6-2}$$

$$\text{RF}(t) = \cos (\omega_i t + \theta_1) - \frac{1}{3} \cos 3 (\omega_i t + \theta_1) + \frac{1}{5} \cos 5 (\omega_i t + \theta_1) \tag{6-3}$$

To keep our approximation from becoming overly complicated, we shall delete any product term where the fractional product is less than 0.1. The product of LO(t) $\times$ RF(t) is given by

$$\mathrm{LO}(t) \times \mathrm{RF}(t) = (\cos \omega_i t) \cos (\omega_i t + \theta_1)$$
$$- \frac{1}{3} (\cos \omega_i t) \cos (3\omega_i t + 3\theta_1) + \frac{1}{5} (\cos \omega_i t) \cos (5\omega_i t + 5\theta_1)$$
$$- \frac{1}{3} (\cos 3\omega_i t) \cos (\omega_i t + \theta_1) + \frac{1}{9} \cos (3\omega_i t) \cos (3\omega_i t + 3\theta_1)$$
$$+ \frac{1}{5} \cos (5\omega_i t) \cos (\omega_i t + \theta_1) \tag{6-4}$$

From fundamental trigonometric identities,

$$\cos \alpha \cos \beta = \frac{1}{2} [\cos (\alpha - \beta) + \cos (\alpha + \beta)] \tag{6-5}$$

Thus the only term that contains a DC voltage as a function of phase (θ_1) is the first term, where

$$(\cos \omega_i t) \cos (\omega_i t + \theta_1) = -\frac{1}{2} \cos \theta_1 = \frac{1}{2} \cos (-\theta_1) \tag{6-6}$$

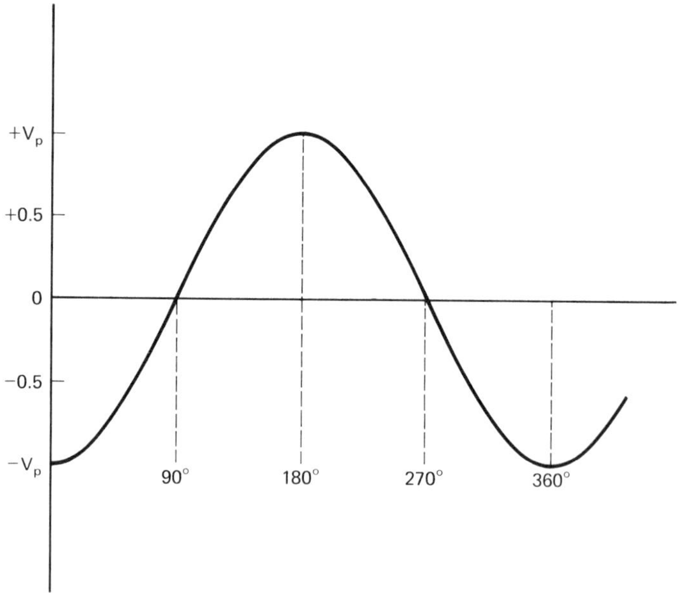

Figure 6-1. Double balanced mixer's phase detector performance

It should be noted that if all terms of the Fourier series had been included, the result would not have changed; therefore, our previous approximation was adequate.

The graph cos $(-\theta)$ for phase differences extending from 0° to 360° is illustrated in Fig. 6-1. Maximum sensitivity, in terms of $\Delta V/\Delta\theta$, occurs at 90° and 270° for this type of phase detector. As the phase difference departs from either 90° or 270° by as much as $\pm 30°$, the sensitivity (or transfer function) will degrade by 5% and would not cause serious problems in most phase-locked loops.

DIGITAL CIRCUITS AS PHASE DETECTORS

The transfer function of a double-balanced mixer, for output voltage as a function of input phase difference, is the cosine of phase difference and is symmetrical about 180°. The digital circuit most similar to the DBM is the exclusive NOR, and this triangular function (Fig. 6-2) is also symmetrical. Sensitivity in units of "volts per degree" is very linear and does not require operating the phase-locked loop at either 90° or 270°.

The sawtooth phase detector with either edge triggering or impulse triggering is obtained from a flip-flop and has a sawtooth characteristic [2] as shown in Fig. 6-3. This type of detector was initially very popular due to the linear voltage versus phase relationship over a complete cycle of

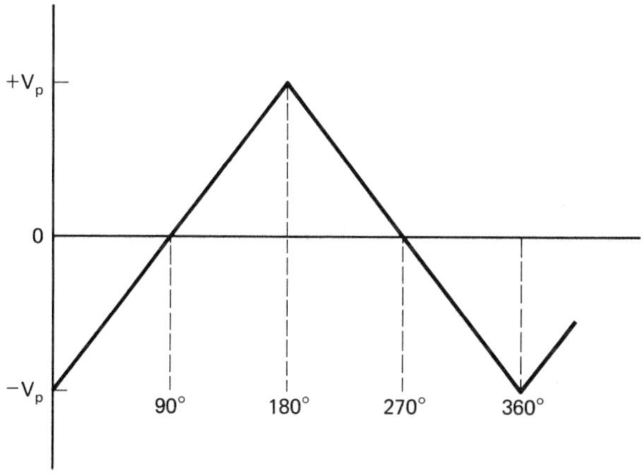

Figure 6-2. Exclusive NOR's phase detector performance

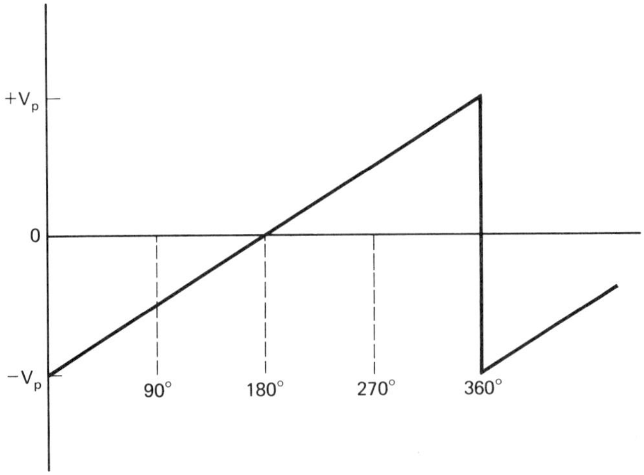

Figure 6-3. Edge triggered flip-flop's phase detector performance

360° without linear slope reversal. This device has been largely replaced with phase-frequency detectors, discussed in Chapter 8.

ELECTRONICALLY TUNED OSCILLATORS

Oscillators needed for specific applications can be separated into two basic types. Microwave frequency synthesizers use either varactor tuning or YIG (yttrium-iron-garnet) tuning to make the oscillator's frequency electronically tunable. The differences between these two types of oscillators are very pronounced and are listed in Table 6-1.

Clearly, when the synthesizer to be designed requires either fast tuning or very low power consumption, the only choice is the varactor-tuned oscillator. When linearity or large bandwidth is the major concern,

TABLE 6-1

	Varactor	YIG
Tuning speed	nanosecond	millisecond
Tuning power	microwatt	watt
Tuning linearity	2% to 10%	< 2%
Tuning bandwidth	octave	multioctave
Loaded Q	5 to 50	500 to 5000

the YIG-tuned oscillator is the obvious choice. Since the loaded Q of the YIG oscillator is much greater than that of the varactor type, it will have the lowest-phase noise sidebands; therefore, most instrumentation applications use the YIG device because of its superior spectral purity and tuning linearity.

The dependence of an oscillator's noise sidebands is described for the oscillator model illustrated in Fig. 6-4. On the left side of the vertical dashed line, a wideband noise source containing a negative series resistance is assumed and could be developed from a wideband amplifier with sufficient positive feedback. The noise voltage V_n is in units of volts per $\sqrt{\text{Hz}}$, and its magnitude is on the order of amplifier gain times the amplifier's input noise voltage. The load, on the right side of the vertical dashed line, consists of an LC series-tuned circuit and external resistance R_x. The transfer function for output load voltage V_o versus input noise voltage V_n is given by

$$\frac{V_o}{V_n} = \frac{R_x}{R_x - R_s + j(\omega L - 1/\omega C)} \tag{6-7}$$

For the condition where the difference of $R_x - R_s$ approaches 0 and $(\omega L - 1/\omega C)$ is much greater than $R_x - R_s$,

$$\frac{V_o}{V_n} = \frac{R_x \omega C}{j(\omega^2 LC - 1)} \tag{6-8}$$

let

$$\omega_o = 1/\sqrt{LC} \tag{6-9}$$

and

$$\omega = \omega_o + \Delta\omega \tag{6-10}$$

where $\Delta\omega$ may assume either positive or negative values.

Next we substitute Eq. (6-10) into Eq. (6-8) for the magnitude of

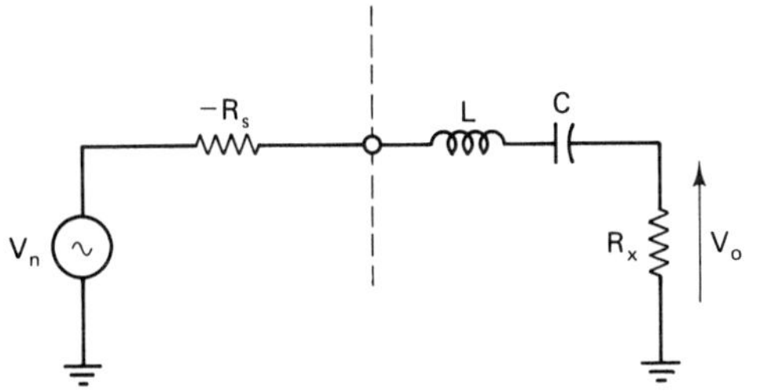

Figure 6-4. Oscillator noise sideband model

V_o/V_n:

$$\left|\frac{V_o}{V_n}\right| = \frac{R_x C(\omega_o + \Delta\omega)}{(\omega^2 + 2\omega_o\Delta\omega + \Delta\omega^2)LC - 1} \tag{6-11}$$

then substitute Eq. (6-9) into Eq. (6-11):

$$\left|\frac{V_o}{V_n}\right| = \frac{R_x(\omega_o + \Delta\omega)}{(2\omega_o + \Delta\omega)L\Delta\omega} \tag{6-12}$$

for $\omega_o >> \Delta\omega$

$$\left|\frac{V_o}{V_n}\right| = \frac{R_x}{2L\Delta\omega} \tag{6-13}$$

As $\Delta\omega$ is increased, the magnitude of V_o/V_n decreases, and the specific value of $\Delta\omega$ that causes $V_o/V_n = 1$ is noted as $\Delta\omega_1$. In Eq. (6-13) we let V_o/V_n equal 1 and derive

$$\Delta\omega_1 = \frac{R_x}{2L} \tag{6-14}$$

We define external circuit Q as Q_x, and it is given by

$$Q_x = \frac{\omega_o L}{R_x} \tag{6-15}$$

where ω_o was defined in Eq. (6-9).

Since bandwidth, BW, is equal to ω_o/Q_x,

$$\Delta\omega_1 = \frac{R_x}{2L} = \frac{\omega_o L/Q_x}{2L} = \frac{\omega_o}{2Q_x} = \frac{BW}{2} \tag{6-15}$$

The relationship between V_o and V_n is shown graphically in Fig. 6-5, where the output voltage limits at the value of V_s as $\Delta\omega$ approaches 0. As $\Delta\omega$ is increased, V_o decreases and is equal to V_n when $\Delta\omega$ is equal to BW/2. The ratio of V_s/V_o as a function of $\Delta\omega$ can only be increased by any of the following three parameters:

1. Increase the oscillator's saturated output voltage V_s.
2. Decrease the value of V_n.
3. Decrease BW/2 by increasing Q_x.

Since V_s and V_n are interrelated because V_n will increase as V_s is increased by using higher power capability devices, increasing Q_x is the better choice for improving an oscillator's spectral purity by lowering its sideband noise voltage content.

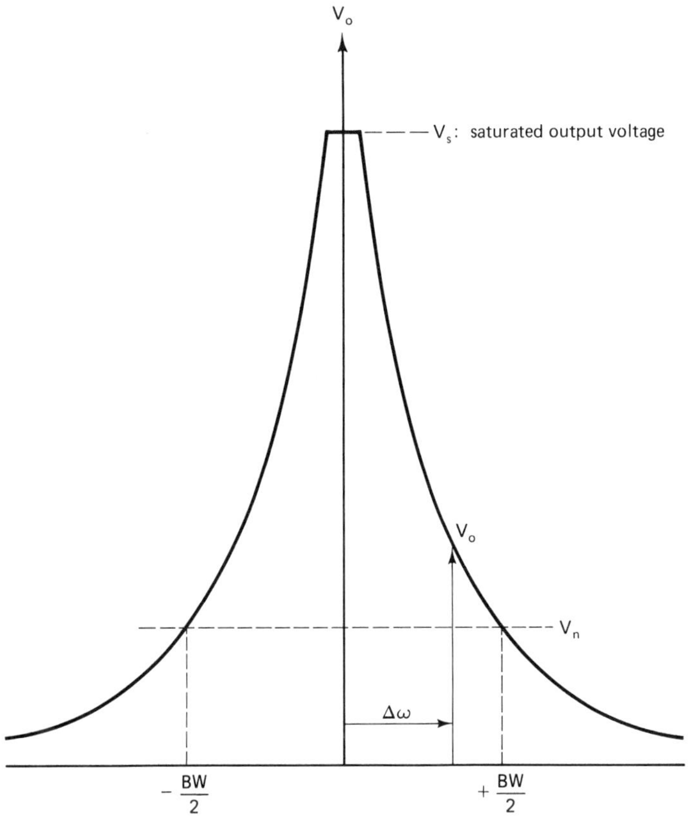

Figure 6-5. Graph of oscillator sideband noise voltage

Generally, in varactor-tuned oscillators, Q_x is in the range of 5 to 50, while in YIG-tuned oscillators, Q_x is usually greater than 1000. If we used the same device to generate V_n and negative resistance R_s and measured V_o with varactor and YIG tuning circuits, the value of BW/2 would be reduced by a factor of 20 or larger with the YIG-tuned circuit.

REFERENCES

1. Kurtz, S. R., "Mixers as Phase Detectors," *Tech-Notes, Watkins-Johnson Co.*, 5, no. 1 (February 1978), 2–15.
2. Hager, J. C., Jr., "Edge-triggered Flip-flops Make 360° Phase Meter," *Electronics* (August 21, 1975), p. 100.

7

The Analog Phase-locked Loop

SINGLE-LOOP TRACKING OSCILLATORS

In frequency synthesizers, the major application of the phase-locked loop (PLL) is to make a voltage-tuned oscillator track a reference frequency standard combined with feedback-programmable frequency dividers and downconverters. Those techniques for demodulation of information or tracking a source in a low signal-to-noise ratio (SNR) environment are not needed in the synthesizer; therefore, we concern ourselves with basic PLL properties. At the end of this chapter, five references are provided on the subject of phase-locked loops.

In this chapter, the analysis deals mainly with how well a phase-locked oscillator will track a reference oscillator's frequency and its spectral sideband content. In Fig. 7-1, a single-loop tracking oscillator is diagrammed for an output frequency f_o and an input reference frequency f_r.

The output from the phase detector, actually a phase-difference detector, is applied to an amplifier and a loop filter where the reference frequency f_r is significantly attenuated.

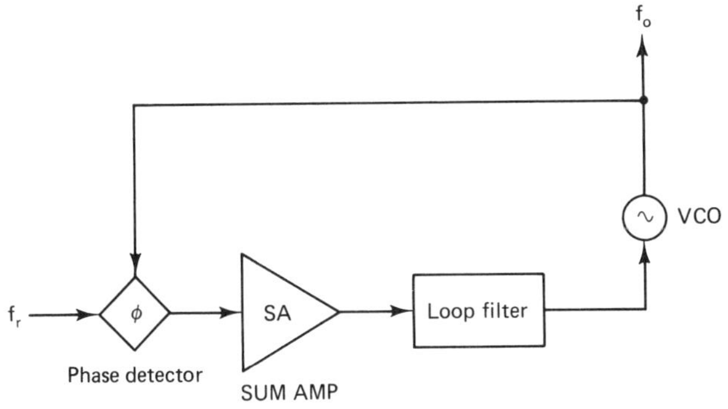

Figure 7-1. Single-loop tracking oscillator

A block diagram model of the loop is illustrated in Fig. 7-2. For the model, the following definitions apply:

1. K_v is the oscillator transfer function in terms of hertz per volt.
2. V_{no}, with the loop opened, represents the equivalent oscillator input noise voltage which causes f_o to have phase noise sidebands and not be an ideal spectral line of infinitesimal width at frequency f_o.
3. The loop filter transfer function is equal to $H(S)$ and is not required to be a function of offset frequency within the loop BW.
4. V_{na} is an added noise voltage, as shown in the loop, caused by intrinsic

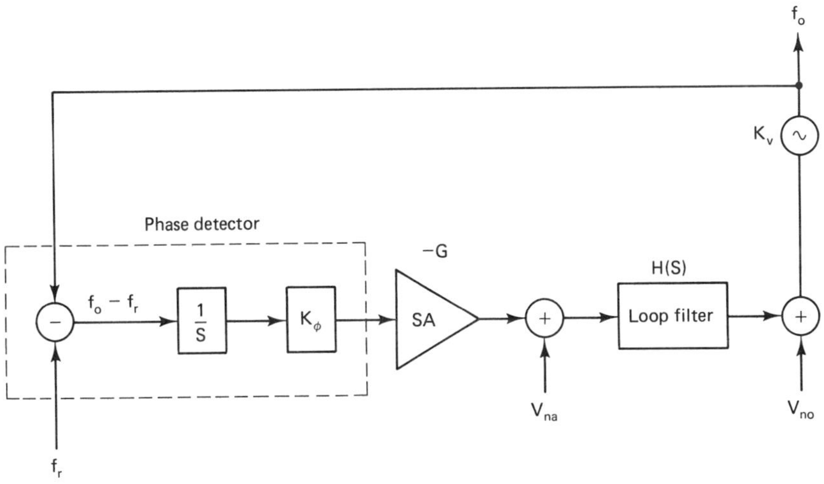

Figure 7-2. Model of the single-loop tracking oscillator

sum amp input noise voltage and/or intrinsic phase detector output noise voltage.

5. The phase detector is modeled within the dashed lines where K_ϕ is in volts per radian and $S = j\omega$.

For the loop, the input-output relation is given by

$$f_o = \left(\frac{K_v}{1 + A}\right)V_{no} + \left(\frac{H(S)K_v}{1 + A}\right)V_{na} + \left(\frac{A}{1 + A}\right)f_r \tag{7-1}$$

where

$$|A| = \frac{GH(S)K_vK_\phi}{f_m} \quad \text{and} \quad A = \frac{GH(S)K_vK_\phi}{jf_m}$$

and is frequency-dependent.

At low-modulation (f_m) frequencies where $|A| >> 1$,

$$f_o = \frac{K_vV_{no}}{A} + \frac{H(S)K_vV_{na}}{A} + f_r \tag{7-2}$$

When the loop is not closed, the free-running oscillator noise sidebands are determined by the product K_vV_{no}. However, in the closed loop, the free-running oscillator noise sidebands are reduced by the magnitude of |A|, and the amount of the reduction, in dB, is given by 20 log |A|.

Added noise sideband signal-to-noise ratio caused by the other noise source V_{na} can be computed from the following equation for each value of f_m:

$$\text{SNR (dB)} = 20 \log \left(\frac{\sqrt{2}\, f_m}{\Delta f}\right) \tag{7-3}$$

where $\Delta f = H(S)(K_v/A)V_{na}$, the rms deviation, and f_m is the modulation frequency (offset or sideband frequency). For $|A| >> 1$, the output frequency will be equal to and track the reference frequency f_r if the changes in f_r occur slowly to maintain the requirement for $|A| >> 1$. In addition, intrinsic oscillator frequency–related sidebands will be reduced as previously noted; however, it is possible to add noise sidebands if the phase detector output or loop amplifier input have high intrinsic noise levels.

Returning to Eq. (7-1):

$$f_o = \left(\frac{K_v}{1 + A}\right)V_{no} + \left(\frac{H(S)K_v}{1 + A}\right)V_{na} + \left(\frac{A}{1 + A}\right)f_r \tag{7-1}$$

we let $|A| << 1$ and obtain Eq. (7-4):

$$f_o = K_vV_{no} + H(S)K_vV_{na} \tag{7-4}$$

and this equation defines the sideband content of f_o at frequencies removed much greater than the loop bandwidth.

The effects of noise sources V_{no} and V_{na} are not reduced except for the filtering of $H(S)$ on V_{na}, and tracking of f_r sideband content cannot be accomplished. Thus the phase-locked loop in our model acts as a high-pass filter to the internal noise sources V_{no} and V_{na} and a low-pass filter to the sideband content of reference f_r.

From Eq. (7-1), where the magnitude of $|A|$ is given by

$$|A| = \frac{GH(S)K_v K_\phi}{f_m} \qquad (7\text{-}5)$$

and if we let f_c be that value of f_m where $|A| = 1$, then

$$f_c = GH(S)K_v K_\phi \qquad (7\text{-}6)$$

If we now let $|GH(S)| = 1$,

$$f_c = K_v K_\phi \qquad (7\text{-}7)$$

and is the approximate loop BW when the phase margin is adequate to have no more than 1 to 3 dB of peaking in the closed-loop gain of $A/(1 + A)$ with respect to f_r.

Usually, the value of $GH(S) \neq 1$, and either $G(S)$ or $H(S)$ is a function as shown in Fig. 7-3. Early PLL designs employed the lag-lead filter, in the top half of the figure, with the gain of the summing amplifier constant to at least a frequency considerably greater than the loop bandwidth. More recent designs favor the circuits shown in the lower half of the figure, and $H(S) = 1$ for this type of PLL. In either case, the value of $|A|$ at a frequency removed (f_m) and less than f_c is the amount of intrinsic oscillator noise reduction that can be obtained. When the primary objective of the PLL design is to have a locked oscillator with minimum-noise sidebands, the loop BW is made equal to the unique value of f_m where the oscillator noise sideband SNR is equal to reference frequency (f_r) noise sideband SNR. If frequency set-on time, or minimum time required to phase-lock the oscillator, is the primary goal, the loop BW is made as large as possible without causing excessive closed-loop gain peaking.

ANALOG FREQUENCY SYNTHESIZER

The tracking oscillator can be configured to follow a multiple of reference frequencies, as illustrated in Fig. 7-4. At the top of the figure, frequency f_o is mixed with a selectable set of frequencies f_h, one at a time, of course,

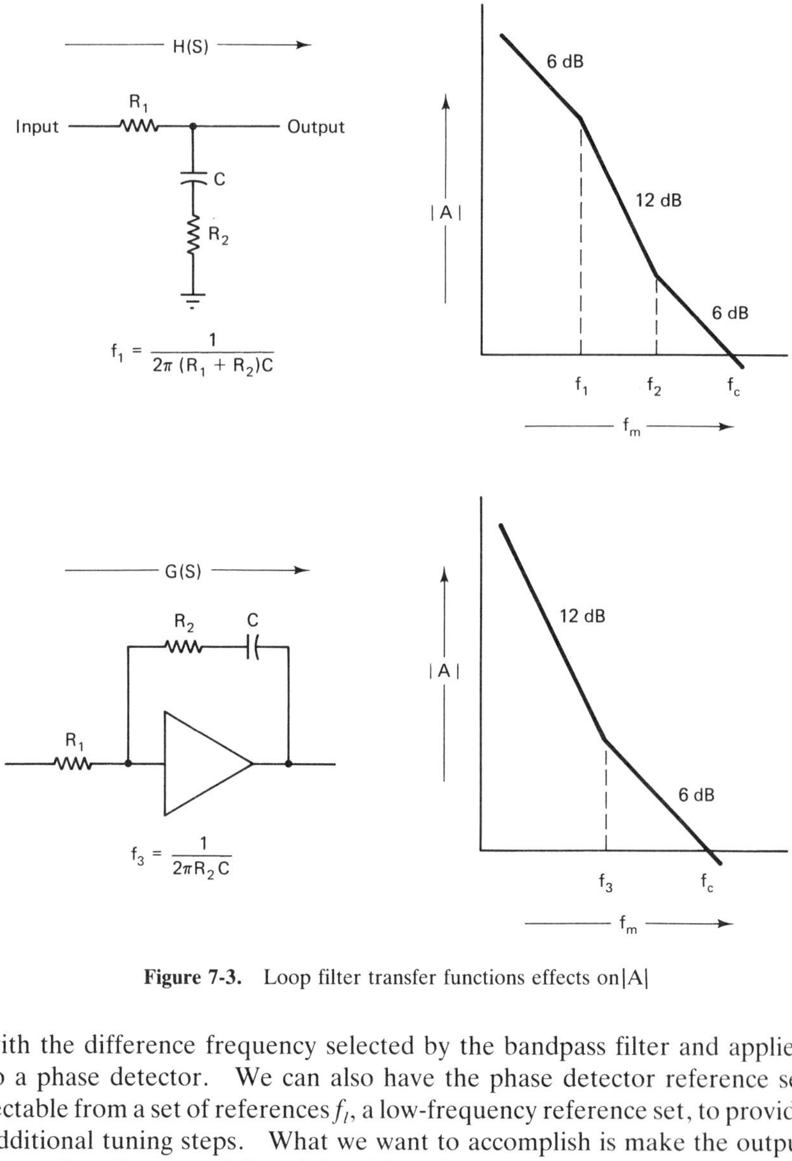

Figure 7-3. Loop filter transfer functions effects on|A|

with the difference frequency selected by the bandpass filter and applied to a phase detector. We can also have the phase detector reference selectable from a set of references f_l, a low-frequency reference set, to provide additional tuning steps. What we want to accomplish is make the output f_o equal to the sum of f_h and f_l. When the loop is phase-locked, the two inputs to the phase detector are equal; therefore,

$$f_o - f_h = f_l$$

The loop can be modeled like Fig. 7-2, and the model of the single-conversion loop is illustrated in Fig. 7-5. We have deleted the noise source V_{na} from this discussion in order to consider only the effects caused by

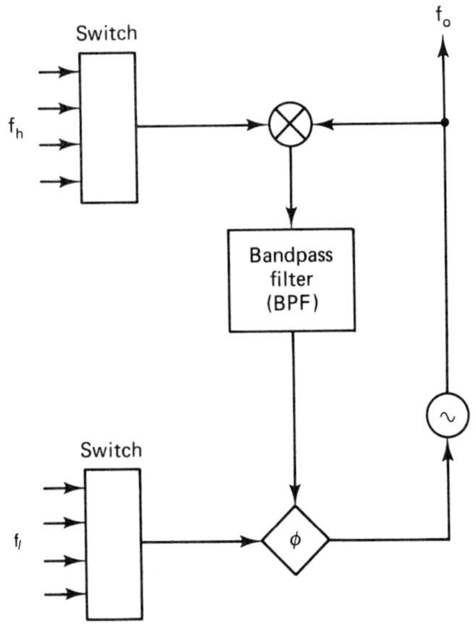

Figure 7-4. Example of analog frequency synthesizer

adding multiple references. In Eq. (7-8), the frequency f_o is determined from

$$f_o = \frac{K_v V_{no}}{1 + A} + \frac{A}{1 + A}(f_n + f_l) \tag{7-8}$$

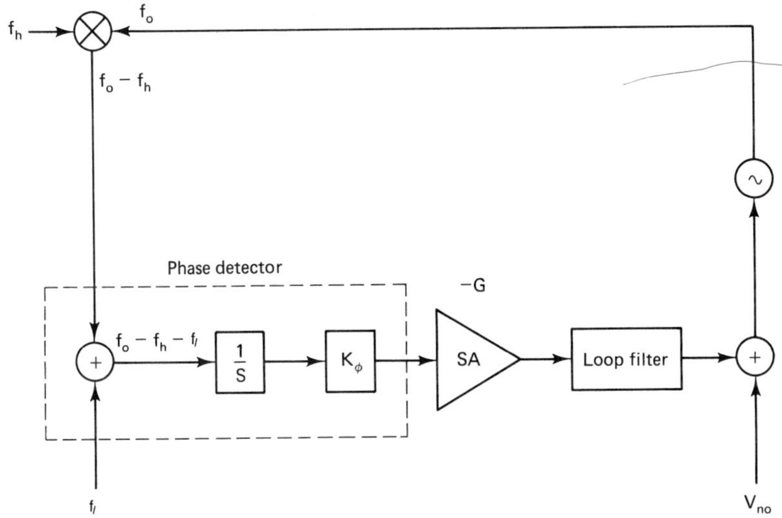

Figure 7-5. Model of the analog frequency synthesizer

where

$$|A| = \frac{GH(S)K_vK_\Phi}{f_m}$$

as before.

From Eq. (7-8) we can make the following observations:

1. Oscillator intrinsic noise represented by V_{no} is reduced exactly as in Fig. 7-2.
2. All spectral components of f_h and f_l within the loop BW (f_c) will be imposed on the output of f_o; thus the reference with the worst close-in spectral purity will determine the spectrum of f_o and mask the lower noise sidebands of the cleaner reference.

In addition, all our concerns for mixer spurious outputs are still valid. Harmonic modulation products of f_o and f_h will be imposed on the power spectrum of f_o if these unwanted spurious frequencies are removed from $(f_o - f_h)$ by no more than an amount equal to f_c, the loop bandwidth. If these spurious frequencies are removed from $(f_o - f_l)$ by an amount much greater than f_c, their effect on f_o will be greatly attenuated.

PULL-IN FREQUENCY

Prior to development of phase-frequency detectors, which are discussed in Chapter 8, this author used the following analysis to calculate the maximum allowable open-loop tuning error to guarantee phase lock. With the model shown in Fig. 7-6, the phase detector is a double-balanced mixer in the unlocked state and has a peak voltage output of V_p from a periodic waveform resembling a sinewave.

With switch S1 in the open position and the oscillator output frequency f_o near the value of the reference f_r, the error frequency f_e is given by

$$f_e = |f_o - f_r| \tag{7-9}$$

The loop may not phase-lock when S1 is closed if the mistuning frequency f_e is large. If we let the open-loop gain be A_0' when the loop is not locked, the open-loop gain is given by

$$A_0' = GH(S) \tag{7-10}$$

where $H(S)$ is the loop filter (composed of R1, C, and R2) attenuation at

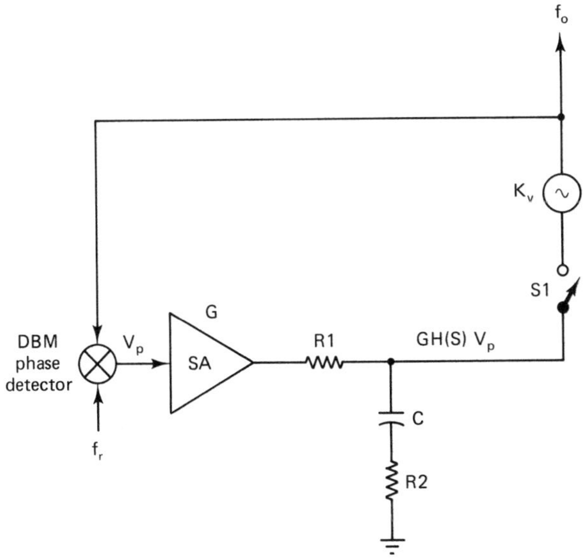

Figure 7-6. Model to estimate the pull-in frequency

error frequency f_e. For the loop always to lock, the peak deviation of f_o must exceed the error f_e if S1 were allowed to close and is given by

$$GH(S)V_pK_v > f_e \tag{7-11}$$

and increasing the loop bandwidth will allow f_e to be larger. This is a safe first-order approximation. Actually, f_e can be slightly larger than $GH(S)V_pK_v$ because the instantaneous beat frequency slows down as f_o approaches f_r and is attenuated by a smaller amount in the loop filter. For these reasons, it was important in the older synthesizer designs to maintain minimum tuning error prior to closing the loop; with phase-frequency detectors, this limitation has been removed.

PLACING DISCRIMINATORS IN THE PHASE-LOCKED LOOP

In this section we establish the advantages of placing a discriminator in a feedback loop around the VCO as an aid in improving the performance of a phase-lock loop. In Fig. 7-7, at the upper left, the VCO transfer function is defined as

$$f_o = f_1 + K_vV_i \tag{7-12}$$

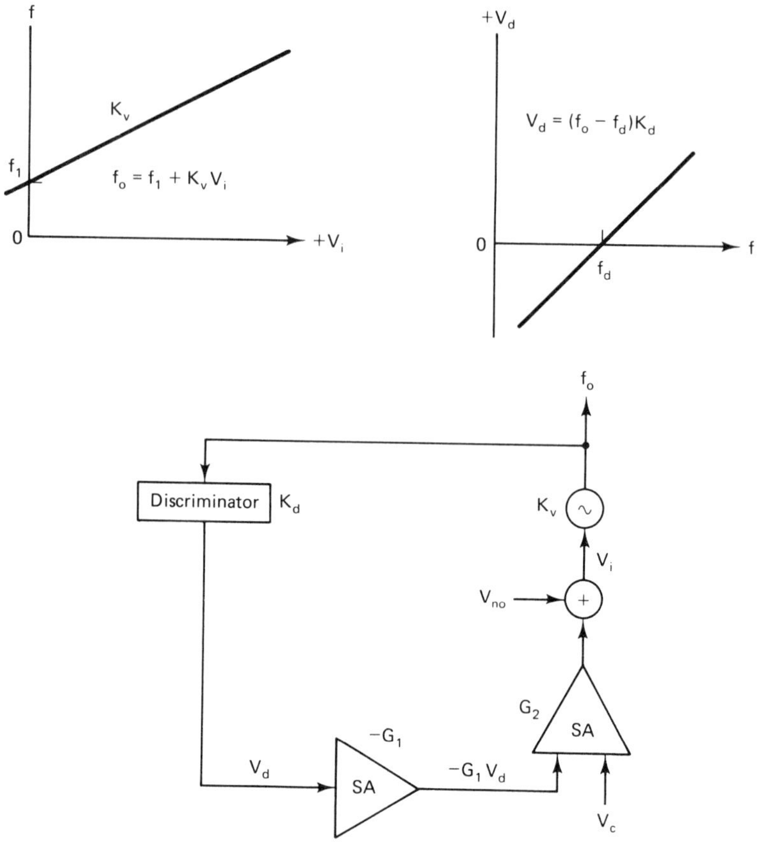

Figure 7-7. Oscillator with discriminator feedback

where f_1 is the output frequency when the input voltage V_i is equal to 0. At the upper right, the discriminator output voltage V_d is given by

$$V_d = (f_o - f_d)K_d \tag{7-13}$$

where the input frequency is f_o, and f_d is that frequency that causes V_d to equal 0 when f_o is equal to f_d.

The model of the discriminator loop is illustrated in the lower half of Fig. 7-7. In this model, V_{no} is the equivalent input noise voltage of the oscillator that causes the output noise sidebands in the open-loop oscillator. The loop contains two summing amplifiers, with gains $-G_1$ and G_2 where an external control voltage V_c is applied to the input amplifier G_2. For this configuration, the output frequency is given by

$$f_o = \frac{f_1}{1 + A} + \left(\frac{K_v}{1 + A}\right)V_{no} + \left(\frac{G_2 K_v}{1 + A}\right)V_c + \left(\frac{A}{1 + A}\right)f_d \tag{7-14}$$

where

$$A = G_1 G_2 K_v K_d \quad (7\text{-}15)$$

Case 1: Open-loop Oscillator

We can open the loop by making $G_1 = 0$ and $G_2 = 1$, and Eq. (7-14) reduces to

$$f_o = f_1 + K_v V_{no} + K_v V_c \quad (7\text{-}16)$$

To have a well-defined relation between f_o and control voltage V_c requires the VCO constant K_v to be very accurate and f_1 to remain constant. This is usually not true since K_v is generally nonlinear and f_1 drifts with time and temperature changes. For these reasons, it is difficult to accurately coarse-tune a varactor-tuned oscillator to well-defined values. In addition, any control voltage noise adds to $K_v V_{no}$ (the intrinsic oscillator open-loop noise) and can degrade spectral purity.

Case 2: Closed Discriminator Loop

By making $G_1 \neq 0$ and $G_2 >> 1$, an effective closed loop can be obtained when $A >> 1$ from Eq. (7-15) and is substituted into Eq. (7-14):

$$f_o = \frac{f_1}{G_1 G_2 K_v K_d} + \frac{K_v V_{no}}{G_1 G_2 K_v K_d} + \frac{V_c}{G_1 K_d} + f_d \quad (7\text{-}17)$$

The discriminator loop reduces variations in constant f_1 and noise sidebands $K_v V_{no}$ of the open-loop condition by $G_1 G_2 K_v K_d$. The term $G_1 G_2 K_v K_d$ is dependent on the modulation frequency f_m; however, when f_m is very small so that $G_1 G_2 K_v K_d$ is very large, the output frequency is determined mainly by the control voltage V_c multiplied by $1/G_1 K_d$ and the discriminator constant f_d. In effect, we have exchanged the open-loop tuning characteristics of the VCO for the inverse of the discriminator's voltage-frequency relationship. Usually, the characteristics of microwave discriminators are better defined than varactor-tuned VCOs; therefore, we can set the VCO to a desired frequency with improved accuracy. With the frequency set-on error reduced to a value less than the phase-lock loop bandwidth, the phase-lock loop will lock very fast.

When a discriminator loop is added inside a phase-lock loop, it is important to make the discriminator's loop bandwidth much greater than the PLL bandwidth. This means at the value of f_m where the PLL gain is 1 and gain is decreasing as f_m is increased, the closed-loop discriminator and VCO combination should appear frequency-independent to the PLL and present a transfer function equal to $1/G_1K_d$. For these conditions, the VCO noise sidebands will be reduced by the magnitude of $|1 + G_1G_2K_vK_d|$ until larger values of f_m cause this term to equal 1. This type of discriminator loop is most effective when the input frequencies to the phase detector are greater than a few hundred megahertz where digital phase-frequency detectors cannot be employed.

REFERENCES

1. McAleer, H. T., "A New Look at the Phase-locked Oscillator," *Proceedings of the IRE* (June 1959), pp. 1137–1141.
2. Rey, T. J., "Automatic Phase Control: Theory and Design," *Proceedings of the IRE* (October 1960), pp. 1760–71.
3. Greenstein, L. J., "Phase-locked Loop Pull-in Frequency," *IEEE Transactions on Communications*, COM-22, no. 8 (August 1974), 1005–13.
4. Gupta, S. C., "Phase-locked Loops," *Proceedings of the IEEE*, (February 1975), 291–306.
5. Govindarajan, M. S., and Rao, B. V., "On a Closed Form Expression for Phase-locked Loop Pull-in Range," *IEEE Transactions on Communications*, COM-24, no. 8 (August 1976), 910–13.

8

Digital Phase-locked Loops

PHASE-FREQUENCY DETECTORS

One of the more important advances in phase-lock loop capabilities was the development of phase-frequency detectors. These frequency and phase comparison circuits have been constructed from a variety of digital circuits including bistable multivibrators, Schmitt triggers, monostable multivibrators, and gates [1]. Some of the early circuits were linear over a range of 2π radians [2]. More recent types fabricated from digital integrated circuits have an input-output transfer characteristic as illustrated in Fig. 8-1 [3].

The major advantage of the phase-frequency detector over the phase comparators described in Chapter 6 is that the pull-in range can be increased to very large values without altering the loop bandwidth; therefore, the pull-in range is essentially independent of loop bandwidth and not limited by loop bandwidth as in the older designs. For these reasons, the remaining chapters describing digital-type frequency synthesizer designs assume that the phase-frequency detector is used in the PLL.

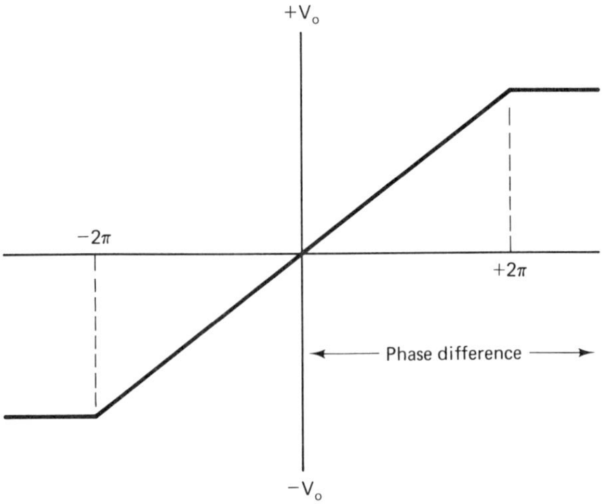

Figure 8-1. Transfer function of phase-frequency detector

SINGLE-LOOP SYNTHESIZER WITH PASSIVE FILTER

Figure 8-2 is a single-loop synthesizer where the output frequency f_o is a function of the reference frequency f_r and the programmable counter N, where N can have a large range of integers usually controlled in minimum step size of unity. This loop is identical to that in Fig. 7-2 except for the addition of counter N and the definition of the noise voltage V_{na}. In Fig. 8-2, this noise voltage is assumed to be caused by the following voltages:

1. The amplifier input noise, increased by gain G
2. The amplified phase detector output noise
3. The frequency counter's intrinsic noise, which adds phase noise to the frequency of f_o/N; this noise is demodulated by the phase detector and is amplified by the gain G

For this loop, the input-output relation is given by

$$f_o = \left(\frac{K_v}{1 + A/N}\right)V_{no} + \left(\frac{H(S)K_v}{1 + A/N}\right)V_{na} + \left(\frac{A}{1 + A/N}\right)f_r \qquad (8\text{-}1)$$

where

$$A = \frac{GH(S)K_v K_\phi}{jf_m}$$

and is frequency-dependent.

Equation (8-1) is identical to Eq. (7-1) of Chapter 7, except that the open-loop gain term A is divided by the value of N in all three denominators. At lower modulation frequency (f_m) where $|A/N| >> 1$,

$$f_o = \left(\frac{NK_v}{A}\right)V_{no} + \left(\frac{NH(S)K_v}{A}\right)V_{na} + Nf_r \qquad (8\text{-}2)$$

Comparing Eq. (8-2) with Eq. (7-2) of Chapter 7 reveals the following:

1. Noise modulation of the oscillator caused by V_{no} is given by $N(K_v/A)V_{no}$ and is larger than without the divider N.
2. Noise modulation caused by V_{na} is given by $N[H(S)K_v/A]V_{na}$ and is larger than without the divider N.
3. Changes in reference frequency (f_r) or its phase are multiplied by N; actually, this configuration with frequency divisor N is a programmable multiplier with respect to reference frequency f_r.

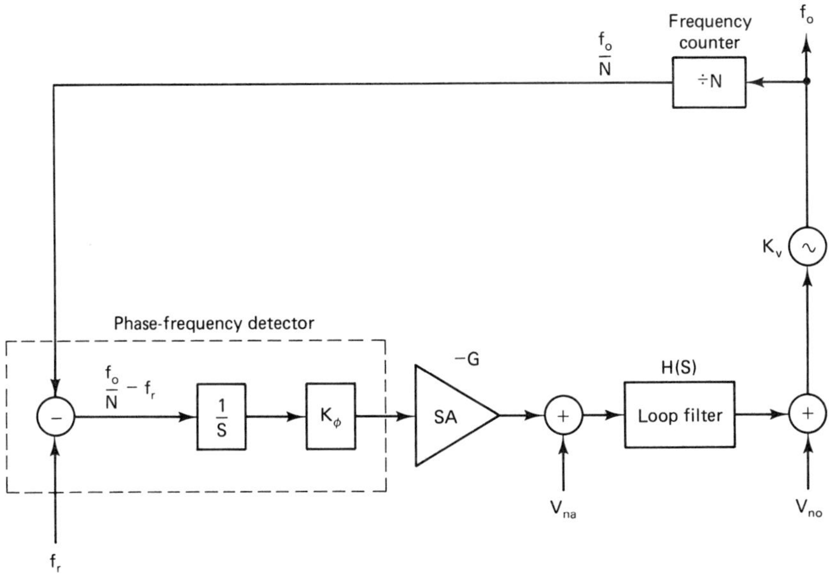

Figure 8-2. Loop model containing a programmable counter

SINGLE-LOOP SYNTHESIZER WITH ACTIVE FILTER

The block diagram model of a more commonly used single-loop frequency synthesizer is presented in Fig. 8-3. In this model, a VCO with transfer constant K_v is controlled by reference frequency f_r and the numerical value of programmable counter N. As before, we let V_{no} represent the noise voltage internal to the oscillator that causes the output frequency f_o to have phase noise sidebands when the oscillator is free-running. The noise voltage V_{ai} is the series equivalent input noise of the amplifier, which has an open-loop gain of $-G$. The noise voltage V_{do} represents the sum of two potential noise sources: One is the intrinsic counter's output noise floor, demodulated by the phase-frequency detector; the other is the intrinsic phase-frequency detector's output noise.

The equations to be discussed in the remainder of this chapter are for the following conditions:

1. $f_1 = \dfrac{1}{2\pi R_2 C}$
2. $f_2 = \dfrac{1}{2\pi(R_1 + R_2)C}$
3. $f_m >> \dfrac{1}{2\pi G R_1 C}$
4. $GR_1 >> (R_1 + R_2)$

The output frequency f_o as a function of loop parameters is given by

$$f_o = \left(\frac{A}{1 + A/N}\right)f_r + \left(\frac{K_v}{1 + A/N}\right)V_{no} - \left[\frac{\left(\dfrac{R_2}{R_1}\right)\left(1 - j\dfrac{f_1}{f_m}\right)K_v}{1 + A/N}\right]V_{do} - \left[\left(\frac{R_1 + R_2}{R_1}\right)\frac{\left(1 - j\dfrac{f_2}{f_m}\right)K_v}{1 + A/N}\right]V_{ai} \tag{8-3}$$

where loop gain A is

$$A = -j\left(\frac{R_2}{R_1}\right)\left(\frac{K_v K_\Phi}{f_m}\right)\left(1 - j\frac{f_1}{f_m}\right) \tag{8-4}$$

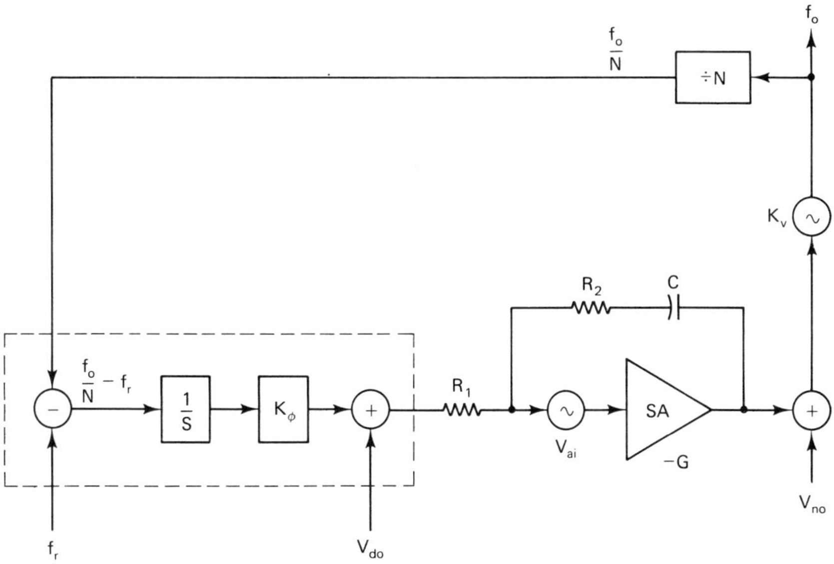

Figure 8-3. Model of single-loop frequency synthesizer

For the loop to be stable, the graph of A/N is usually between the limits as shown in Fig. 8-4. If we let f_3 represent the frequency for $|A/N| = 1$, and if $f_3 > 3f_1$, the approximate value of f_3 is given by

$$f_3 \approx \left(\frac{R_2}{R_1}\right)\left(\frac{K_v K_\phi}{N}\right) \tag{8-5}$$

Equation (8-5) is useful to calculate any single unknown value when all other parameters are defined.

Returning to the general equation (8-3) with the conditions $A/N << 1$, $f_1/f_m << 1$, and $f_2/f_m << 1$ for large values of f_m,

$$f_o = |A|f_r + K_v V_{no} - \left(\frac{R_2}{R_1}\right)K_v V_{do} - \left(\frac{R_1 + R_2}{R_1}\right)K_v V_{ai} \tag{8-6}$$

we note that modulation of f_r is multiplied by $|A|$, and FM noise sidebands caused by V_{no} are not attenuated by loop gain. We now have additional FM noise sidebands caused by noise voltages V_{do} and V_{ai}; therefore, noisy loop components can degrade the spectral purity of the VCO when compared to the free-running, unlocked oscillator's FM spectrum. The unwanted effects of these additional noise sources can be reduced by adding a low-pass filter to the amplifier output; however, care must be taken to prevent unity loop gain phase margin from becoming too small.

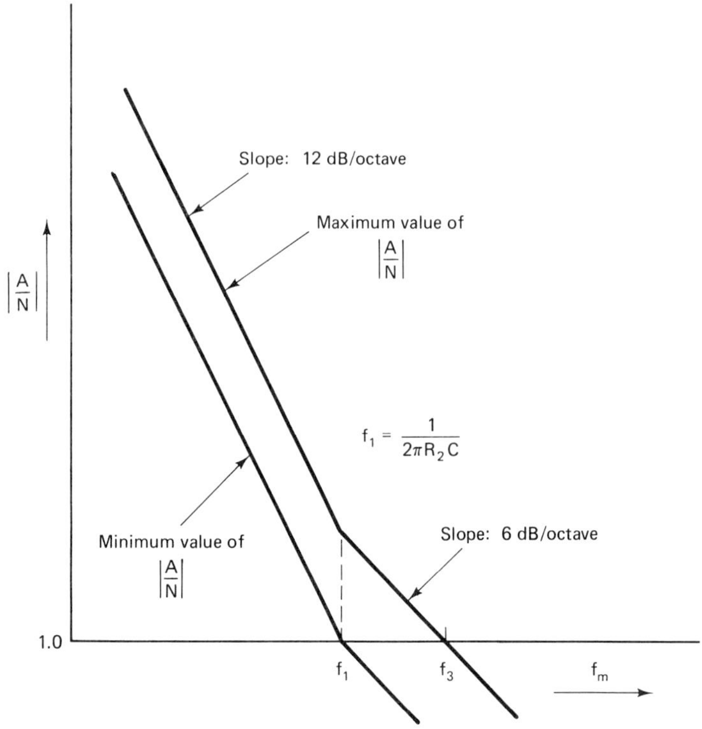

Figure 8-4. Graph of stable loop

When $A/N >> 1$, the general equation reduces to

$$f_o = Nf_r + \frac{K_v V_{no}}{A/N} - \left(\frac{R_2}{R_1}\right)\frac{\left(1 - j\dfrac{f_1}{f_m}\right)}{A/N} K_v V_{do} - \left(\frac{R_1 + R_2}{R_1}\right)\frac{\left(1 - j\dfrac{f_2}{f_m}\right)K_v}{A/N} V_{ai} \tag{8-7}$$

and the following noise sources are reduced:

1. $K_v V_{no}$ is reduced in magnitude by A/N.
2. $\left(\dfrac{R_2}{R_1}\right)K_v V_{do}$ is reduced by A/N but multiplied by $\left(1 - j\dfrac{f_1}{f_m}\right)$.

3. $\left(\frac{R_1 + R_2}{R_1}\right)K_v V_{ai}$ is also reduced by A/N but multiplied by $\left(1 - j\frac{f_2}{f_m}\right)$.

The general equation given by Eq. (8-3) can be a useful design tool for designing loops of this type or evaluating their performance. The negative signs associated with V_{do} and V_{ai} could have been changed to positive values since we are representing noise voltages and it is not important if they are 180° out of phase; however, leaving them negative as originally derived allows the injection of voltages at either V_{do}, V_{ai}, or V_{no} to measure and evaluate loop performance. In Chapters 9 through 13, the emphasis is on design techniques and design examples for a variety of applications. The analysis of loop behavior, in particular the spectral effects of loop noise voltages developed in Chapters 6 through 8, is valid and can be used to predict noise sidebands in these later chapters. For additional details of synthesizers using digital PLLs, see references [4] and [5].

DESIGN EXAMPLES

Let us look at two design examples using the models of Figs. 8-2 and 8-3. The first model, shown in Fig. 8-5, is similar to Fig. 8-2 with the filter function $H(S)$ within the dashed lines. In this model, the low-pass filter is needed to remove reference f_r and its harmonics so that they do not modulate the VCO; in addition, it allows the average phase detector output voltage to be an input to the summing amplifier. For the low-pass filter not to affect the gain and phase margins of the loop, its 3-dB cutoff frequency is assumed to be much larger than unity gain of the closed loop.

In this example we assume that the output frequency is to be programmable from 50 to 99.9 MHz with 100-kHz step size. From the last term in Eq. (8-2), reference f_r must equal 100 kHz and divisor N is required to be programmable from 500 to 999. To make this example more realistic, we assign the following parameters to the VCO:

1. At input voltage V_i of -3 VDC, f_o is 50 MHz, and K_v is equal to 10 MHz/V.

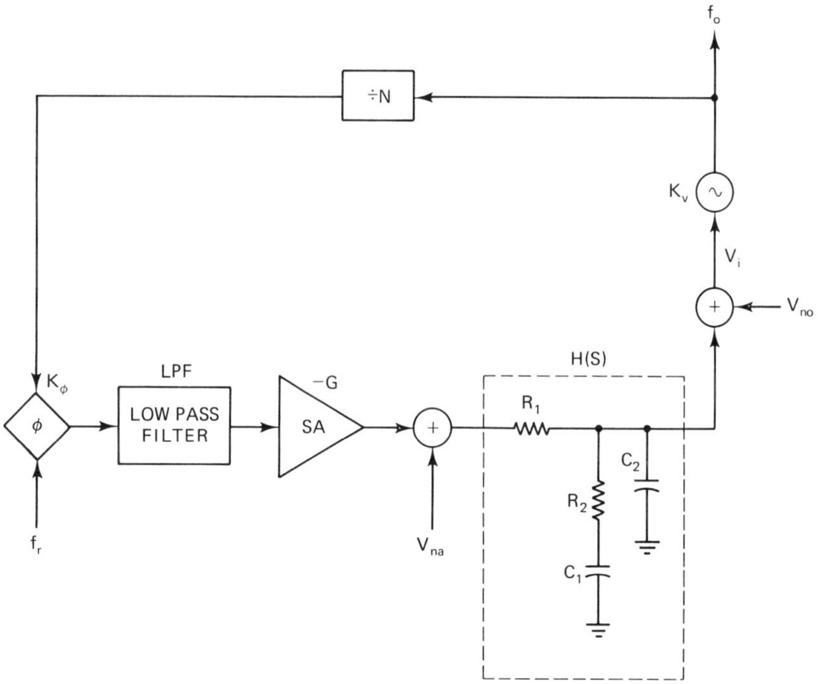

Figure 8-5. Design example no. 1

2. At V_i of -13 VDC, f_o is equal to 100 MHz, and K_v is equal to 3 MHz/V.

For the phase detector, K_ϕ is equal to 5 VDC/4π radians.

The first step in the design processes is to determine the amount of amplifier gain needed to maintain the VCO phase-locked. We need to have adequate gain so that the phase detector output voltage is within the range of minimum and maximum output voltage for this condition. Since most digital phase detectors do not quite reach 0 or 5 V, we select 0.5 V and 4 V as our limits. When the VCO requires -13 VDC, the phase detector output will be less than 4 V, with an amplifier gain of 4.

The second step in the design is to determine the remaining component values in order to assure loop stability. In Eq. (8-1), the open-loop gain term A when divided by N is

$$A = \frac{GH(S)K_vK_\phi}{jNf_m} \tag{8-8}$$

Initially, let $H(S) = 1$ and solve for f_{m1} for $|A| = 1$:

$$f_{m1} = \frac{GK_vK_\phi}{N} \tag{8-9}$$

The variation of f_{m1} is dependent on K_v and N, at the limits for $f_o = 50.0$ MHz, $K_v - 10$ MHz/V, and $N = 500$:

$$f_{m1} = \frac{4 \times 10 \times 10^6 \times 5/4\pi}{500} \approx 31.5 \text{ kHz}$$

for $f_o = 99.9$ MHz, $K_v = 3$ MHz/V, and $N = 999$:

$$f_{m2} = \frac{4 \times 3 \times 10^6 \times 5/4\pi}{999} \approx 4.7 \text{ kHz}$$

Figure 8-6 is a graph of 20 log $|A/N|$ where $f_{m1} = 31.5$ kHz and $f_{m2} = 4.7$ kHz. This is not a stable loop because f_{m1} is too large and greatly exceeds $f_r/20$. Since the maximum value of f_{m1} is 31.5 kHz, the loop gain needs to be reduced by a factor of 6 with the lag-lead composed of R_1, R_2,

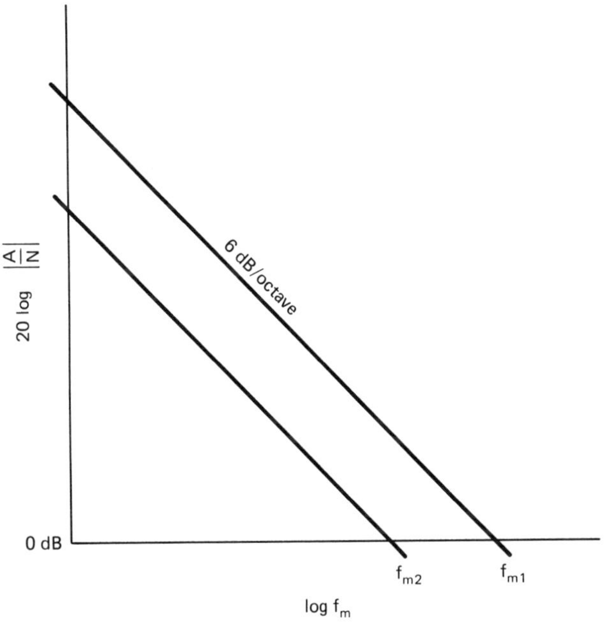

Figure 8-6. Graph of open loop gain. A/N, example no. 1

and capacitor C. The final lag frequency determined by R_2 and C_2 is made greater than f_{m1} for attenuation of V_{na}. This requirement for 5 kHz maximum BW is obtained by making the ratio of R_1 to R_2 equal to 6. The lag frequency given by $1/2\pi\ (R_1 + R_2)C$ needs to be less than 4.7 kHz/6 to make the loop stable when the lower limit of $f_{m2}/6$ is approached.

The intent of this example is to illustrate a few of the loop characteristics when there are large variations in K_v and N. To make f_{m1} and f_{m2} nearly equal for all values of N ranging from 500 to 999 requires that some form of linearization be added to the loop. This can be achieved by increasing the amplifier gain G as N is increased. To estimate the SSB phase noise of the locked oscillator, the magnitude of A/N in Fig. 8-6 is used to estimate the reduction of noise voltages V_{no} and V_{na} in Eq. (8-1).

The second design example, shown in Fig. 8-7, is very similar to Fig. 8-3. For the same values of f_o, K_v, V_i, N, and f_r as in the first example, the first step is to determine the maximum value of f_3 using Eq. (8-5). If we let $R_2/R_1 = 4$, the AC gain of the loop amplifier, we have equations identical to the previous example of Eq. (8-9) where gain G was the closed-loop amplifier DC gain. In Fig. 8-7, amplifier gain G is the open-loop DC gain and is usually very large (> 10,000).

Figure 8-8 is a graph of $20 \log |A/N|$ versus $\log f_m$ with f_3 at its lower and upper limits, corresponding to f_{m2} and f_{m1} in the previous example. From the graph, it is obvious that f_1 must be lower than the lowest value

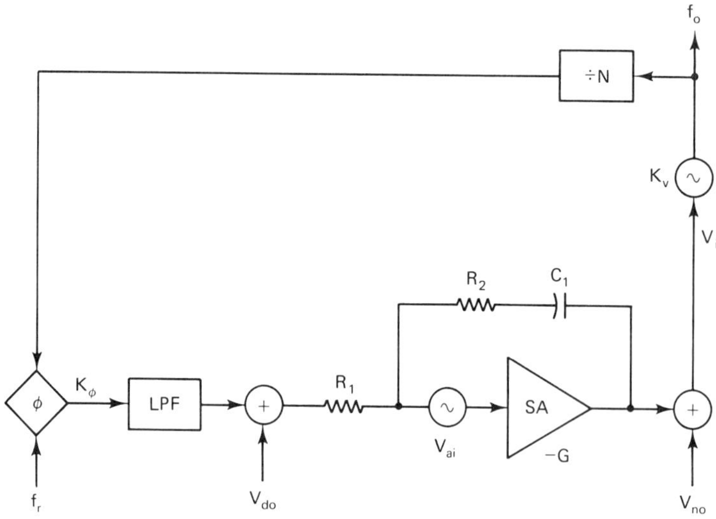

Figure 8-7. Design example no. 2

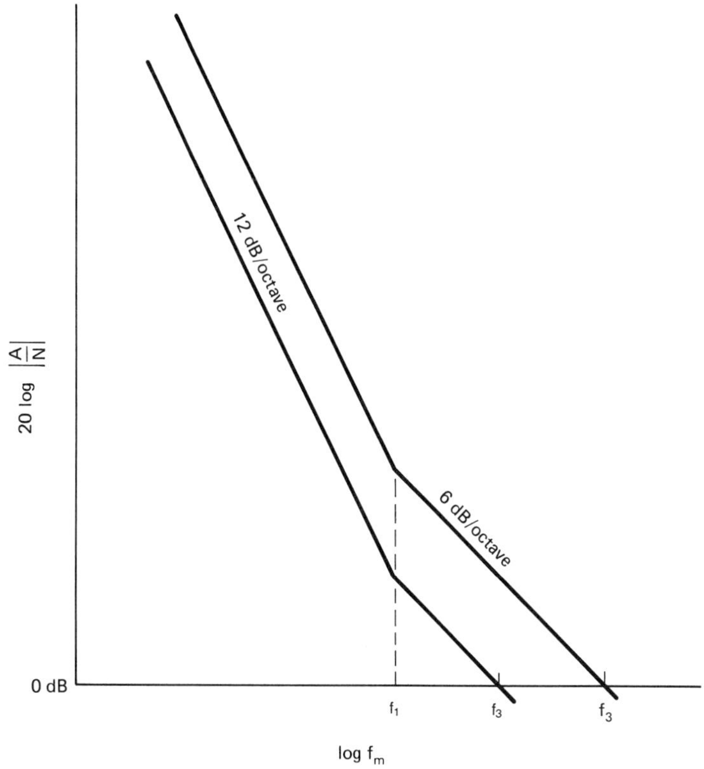

Figure 8-8. Graph of open loop gain A/N, example no. 2

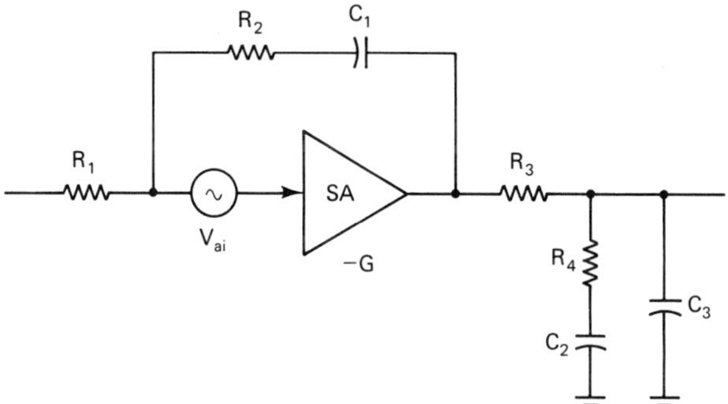

Figure 9-9. Combined filters of examples no. 1 & no. 2

of f_3 to have adequate phase margin in the loop. A good choice is to make $f_1 \leq 1/3$ the lower limit of f_3. This would make f_3, lower limit, divided by 3 equal to $1/2\pi R_2 C$.

Frequently, the filter functions in the two preceding examples can be combined to achieve the largest amount of externally induced noise voltage reduction by combining the loop filters as shown in Fig. 8-9. With the summing amplifier gain G very large, the input frequencies to the phase detector will maintain a nearly constant phase difference of 0 phase if the phase detector output is 0 V when the input phase differential is 0. The filters of examples 1 and 2 can be combined by initially designing the loop as in example 2 for making the loop stable without the addition of R_3, R_4, C_2, and C_3 as shown in Fig. 8-9. The graph of 20 log $|A/N|$ is illustrated in Fig. 8-10 with f_1' substituted for f_1 in Eq. (8-7) and is equivalent to f_1 in Fig. 8-8. We now decrease f_1' to a new value of f_1 as shown in Fig. 8-10.

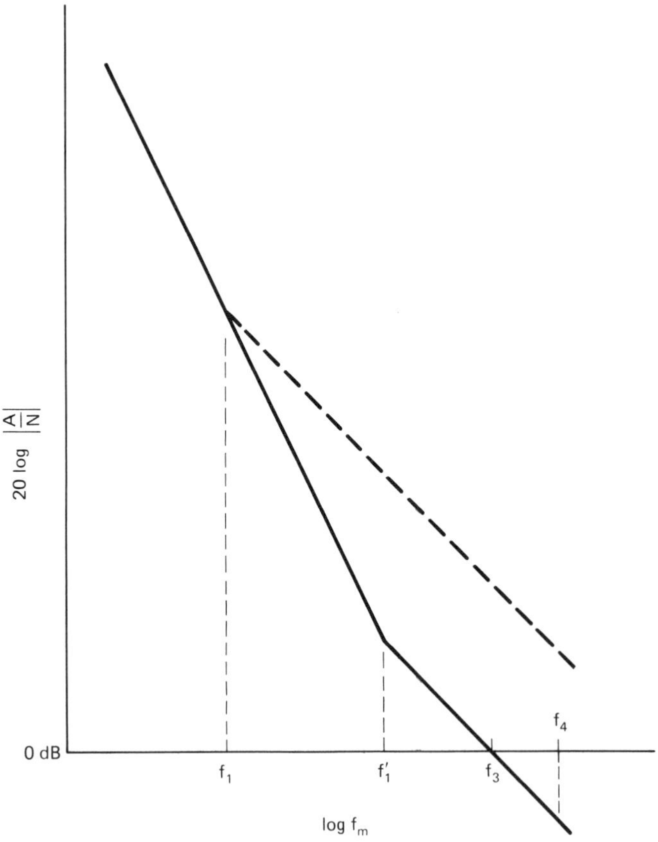

Figure 8-10. Graph of combined loop filter functions

The ratio of f_1'/f_1 is the amount of attenuation added to the loop by R_3, R_4, and C_2. The new value of f_1 is given by

$$f_1 = \frac{1}{2\pi R_2 C_1} = \frac{1}{2\pi(R_3 + R_4)C_2} \tag{8-10}$$

and the value of f_1' is

$$f_1' = \frac{1}{2\pi R_4 C_2} \tag{8-11}$$

and lag f_4 is

$$f_4 = \frac{1}{2\pi R_4 C_3} \tag{8-12}$$

The main advantage to this combination of loop filters is that the lag-lead-lag of R_3, R_4, C_2, and C_3 can be placed within the VCO housing to filter the control voltage applied directly to the VCO's varactor diode.

REFERENCES

1. Brown, J. I., "A Digital Phase and Frequency Detector," *Proceedings of the IEEE* (April 1971), pp. 717–18.
2. Herbst, C. A., "Digital Phase-locked Loop with Loss-of-Lock Monitor," *Electronic Design News* (October 15, 1971), pp. 64–65.
3. Motorola MECL Phase-Frequency Detectors Part Numbers MC4040 and MC12040, 1973.
4. Rohde, U. L., *Digital PLL Frequency Synthesizers*. Englewood Cliffs, N.J.: Prentice-Hall, 1983.
5. Byrne, C. J., "Properties and Design of the Phase-controlled Oscillator with a Sawtooth Comparator," *Bell System Technical Journal* (March 1962), pp. 559–602.

9

Multiple Digital Phase-lock Loops

In this chapter, we begin with an examination of a two-loop synthesizer technique and then expand the concept to include technical approaches where multiple loops can be added to achieve smaller tuning increments [1] without degrading spectral purity or significantly increasing the time required to change frequencies. If, in a single loop, we attempt to have smaller tuning increments by lowering the phase-detector reference f_r and increasing the value of N in the feedback counter, two major detrimental conditions are encountered.

The first effect is that at frequencies removed from the carrier within the loop bandwidth, the phase noise present in the sidebands of reference f_r are multiplied by the larger value of N; thus spectral purity is degraded. These noise sidebands are increased (in dBc) by

$$\text{dBc increase} = 20 \log N$$

For example, if $N = 1000$, this would cause a 60-dB lowering of carrier-to-noise sideband ratio. The second effect is that loop bandwidth must be smaller, and this results in a longer time to lock and settle out the phase transient each time N is programmed to a new value. The major limit to maximum loop bandwidth is the value of reference frequency f_r. Since the counter output frequency is equal to f_r at phase lock, the information rate around the loop has a time delay that is inversely proportional to f_r. The time delay causes a phase shift that is defined by

$$\phi_r = \frac{f_m}{f_r} \times 360°$$

where ϕ_r is a phase shift within the loop that changes linearly with frequency. For example, if f_r were equal to 100 kHz and unity loop gain f_m were designed to be 10 kHz, the phase margin would be reduced by 36°. Since we have 90° phase margin in a PLL without subtracting the phase shift caused by other loop components, the loop phase margin would be reduced to 54°. When another 30° to 45° of phase shift due to loop components is added, the phase margin is reduced to 9° and is not acceptable. To have adequate loop phase margin (as a general rule) requires the loop bandwidth to be less than 1/20 of the phase-detector reference frequency.

TWO-LOOP SYNTHESIZER

The model diagrammed in Fig. 9-1 contains two phase-locked loops. The loop on the right side, labeled "tracking loop," has an output frequency f_o defined by

$$f_o = f_l + f_2 \tag{9-1}$$

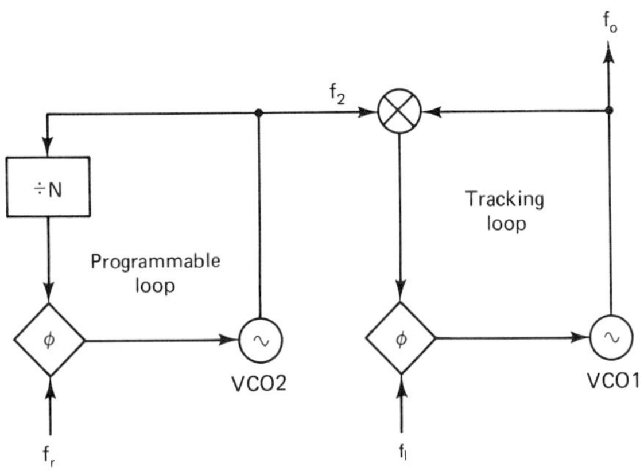

Figure 9-1. Model of two loop synthesizer

where f_l is the lower frequency reference and f_2 is the output of a single programmable loop and is defined by

$$f_2 = Nf_r \tag{9-2}$$

and f_r is a second reference frequency.

Combining these two equations to define output f_o as a function of two reference frequencies yields

$$f_o = f_l + Nf_r \tag{9-3}$$

STABILITY AND SPECTRAL PURITY

To estimate the double-loop performance at the output frequency f_o, we can divide our analysis into two separate parts. First, consider the tracking loop as containing a fixed divider N, where $N = 1$. With this assumption, the analysis of Chapter 8 is valid, or we can use the analysis of Chapter 7 to estimate loop stability and the level of phase-noise sidebands (power spectral density) as a function of reference f_l. Next we consider stability and noise sidebands as a function Nf_r using the analysis of Chapter 8. The composite phase noise from these two separate sources is added using the square root of the sum of the rms values squared since the sources of noise are uncorrelated. Usually, for this type of double loop, the output spectral purity at frequency f_o will be determined mainly by Nf_r. For a different approach to two-loop synthesizers, see reference [2].

MULTIPLE LOOPS

Let us look at two examples of techniques that use a few common circuits in an add-and-divide approach to achieve very small tuning increments in the output tracking loop. The first example is a base 10 design, and its main advantage is that the divided signal outputs can be at higher frequencies and allow very fast tuning from one frequency to another. Figure 9-2 shows a base 10 synthesizer containing two common circuit pairs, which are used throughout the design. These common circuits are a tracking loop and a loop with a programmable divider, 90–99.

Starting at the bottom of Fig. 9-2, the tracking loop containing VCO3 is tuned in steps of unity from 100 to 109 (the frequency value in hertz,

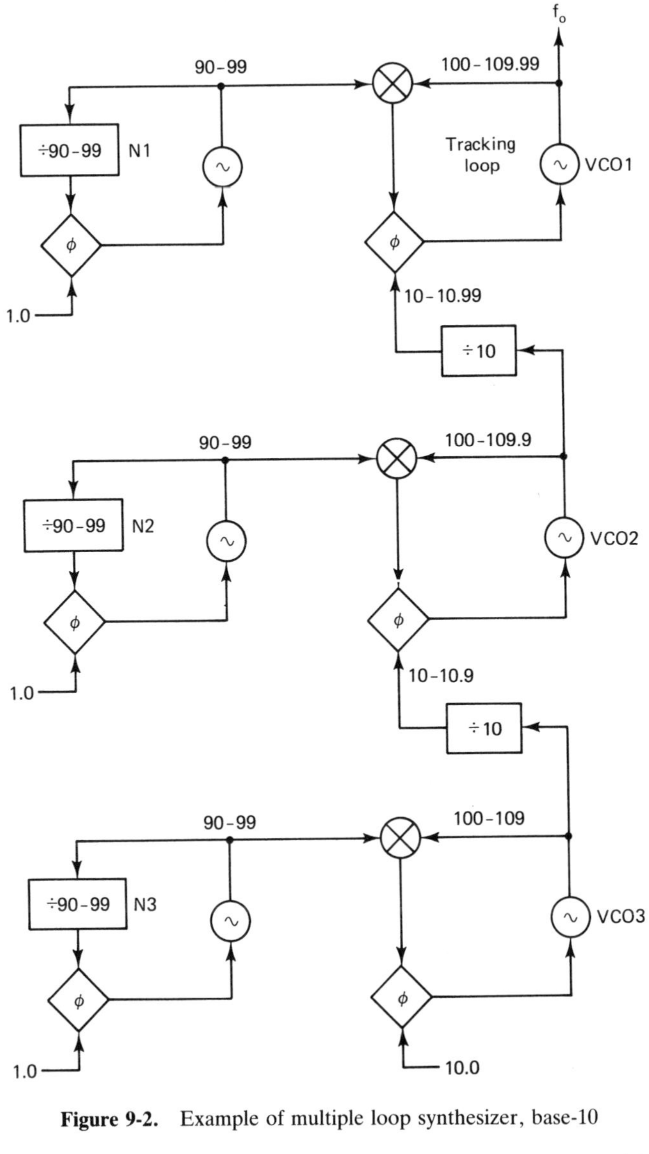

Figure 9-2. Example of multiple loop synthesizer, base-10

kilohertz, or megahertz is not used for this normalized example) by setting the value of N3 in the programmable loop from 90 to 99. Note that this loop pair is identical to the configuration illustrated in Fig. 9-1. The output from VCO3 is divided by 10 prior to being applied to the phase detector as a reference for the middle tracking loop. The divide-by-10 suppresses phase-noise sidebands by 20 log N and is equal to 20 dB; therefore, spectral purity of the 10 to 10.9 variable reference is improved.

At the middle tracking loop, the output frequency is tunable from 100 to 100.9 in tuning steps of 0.1. The lower loop supplies, at VCO2 output, increments from 0 to 0.9, and divider N2 supplies 0 to 9 in unity steps. The output frequency from this tracking loop is also divided by 10 and lowers the phase-noise sidebands from VCO2 by 20 dB and adds another 20 dB of suppression to the phase-noise sidebands associated with VCO3.

At the top of Fig. 9-2, the final output loop pair allows controlling VCO1 from 100 to 109.99 and is governed by

$$f_o = 1 \times N1 + \frac{1}{10}\left[1 \times N2 + \frac{1}{10}(1 \times N3 + 10)\right]$$
$$f_o = 100 + (N1 - 90) + \frac{(N2 - 90)}{10} + \frac{(N3 - 90)}{10^2} \tag{9-4}$$

where N1, N2, and N3 are independently controllable from 90 to 99 in unity increments. If smaller tuning increments are desired, additional loop pairs can be added at the bottom of the block diagram to obtain the desired value.

There are three main advantages to this base 10 synthesizer design:

1. Low design cost because the synthesizer contains two unique circuits
2. Spectral purity dependent mainly on the output loop pair since all the other pairs' phase noise is reduced by at least 20 dB
3. Fast tuning speed since the divider values are not very large, which allows the phase-detector reference frequencies to be at high values

There is one possible problem with this synthesizer approach. If, during a transient condition, an output tracking VCO frequency can fall below 90, the loop may lock with the output VCO much lower in frequency than it should be. To prevent this condition, a frequency comparison circuit can be put in parallel with each mixer and is used to close a switch on the control voltage of the VCO in the tracking loop. With the addition of this feature requiring the tracking-loop VCO to be greater than the left-side VCO before the loop can be closed, the synthesizer is prevented from locking to any false modes.

Tuning speed, the time required to change from one frequency to another, is determined mainly in the loop containing the programmable divider since its phase-detector reference frequency is at most one-tenth the tracking-loop phase-detector reference frequency. Since the programmable-loop VCO has the slower time response, the tracking loop will follow these frequency transients as they occur with very little delay. For this reason, output tuning steps in f_o can be approximated as being equal to the time delays associated with any of the three programmable loops. However, if many loop pairs are in the design, the delay through each tracking loop must be added for the smaller tuning increments.

BASE 100

In the base 10 design, the number of selectable frequencies f_n is given by

$$f_n = 10^N$$

where N is the number of programmable dividers, one per loop pair, in the synthesizer. In our previous example with three dividers, a total of 1000 selectable frequencies was available. If we had a need for 10^6 selectable frequencies, six loop pairs would be required. In the base 100 technique illustrated in Fig. 9-3, we can obtain 10^6 selectable frequencies with just a total of three loop pairs; however, the programmable divider is more complex. This allows a lower-cost synthesizer to be manufactured mainly on the basis of a lower number of total circuits (parts) in the design.

The output frequency f_o of the synthesizer is equivalent to

$$f_o = 100 + (\text{N1} - 990) + \frac{(\text{N2} - 990)}{100} + \frac{(\text{N3} - 990)}{100^2} \qquad (9\text{-}5)$$

where N1, N2, and N3 are independently controlled from 990 to 1089 in unity increments. As with base 10, if smaller tuning increments are needed, we can add a loop pair (for a total of four) and achieve 100 times more selectable frequencies in 1/100 smaller tuning increments than with the three-loop-pair design.

The major disadvantage of base 100 versus base 10 is that set-on frequency delay time is one-tenth the speed since the phase-detector reference frequencies are one-tenth as high and this causes loop bandwidth to be one-tenth as wide. In addition to slower set-on frequency response, phase-noise sidebands will be worse since the divider is approximately ten times larger and degrades spectral purity by 40 dB with respect to the divider noise floor. Also, with a narrow-loop bandwidth, the VCO's intrinsic phase noise is not reduced to the levels that can be obtained in the wider-loop-bandwidth base 10 design. For these two limitations, the fixed divide-by-100 (40 dB) may not be able to improve the output spectral purity of f_o since the loop containing divider N1 will have higher phase-noise sidebands than the output tracking loops' phase-detector reference frequency.

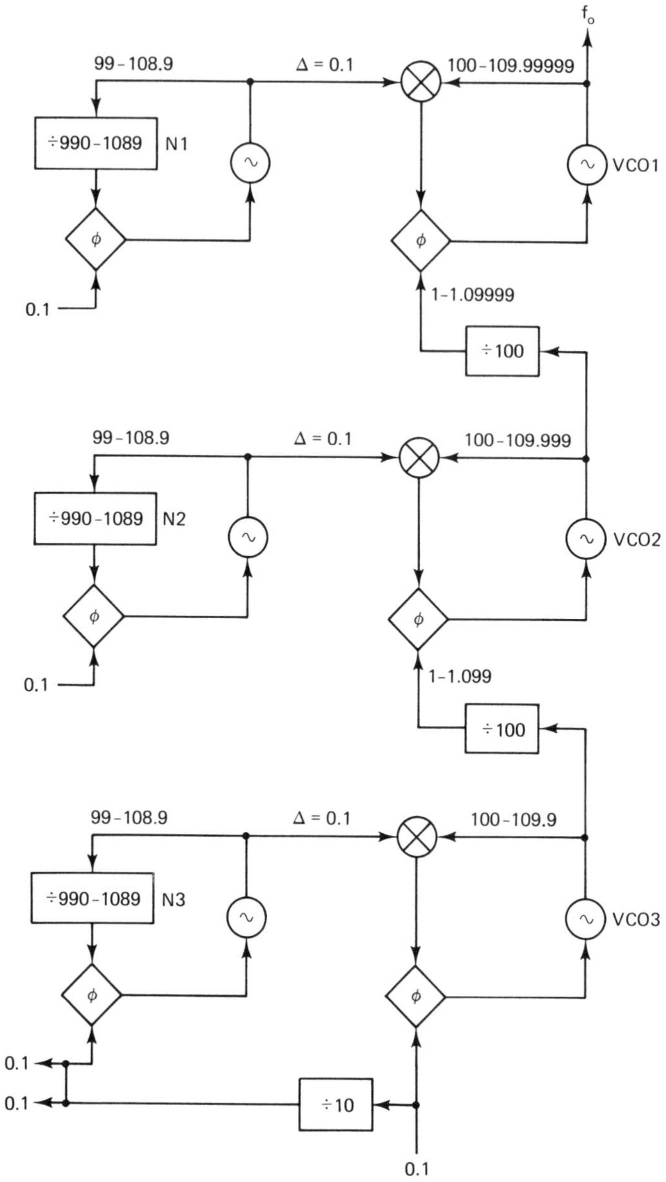

Figure 9-3. Example of multiple loop synthesizer, base-100

Since the input frequencies to the mixer in each loop pair are nearly equal, it is important to ensure that the tracking loop frequency can never be lower than the other input to the mixer during transient conditions when loops are unlocked and are in the processes of relocking to new frequencies. One technique to accomplish this requirement, mentioned earlier, is to add additional circuits to the tracking loop, as shown in Fig. 9-4. The additional circuits are as follows:

1. Two equal frequency dividers ($\div X$)
2. Phase/frequency detector 2
3. Switch driver (SD)
4. Switch S1

For this example, we define *normal loop* tracking when $f_o > f_h$; therefore, when f_o unlocked is less than locked f_o, the mixer output frequency f_i will be less than low-frequency reference f_r to phase/frequency detector 1. The loop is designed to increase the VCO output f_o to a higher value until $f_i \approx f_r$ and then phase-lock can be obtained. When unlocked f_o is greater than locked f_o, frequency f_i is greater than f_r and phase detector 1 causes the output f_o to decrease until $f_i \approx f_o$, when phase locking can be obtained. For these two conditions where f_o is greater than f_h, phase detector 2 causes switch S1 to close.

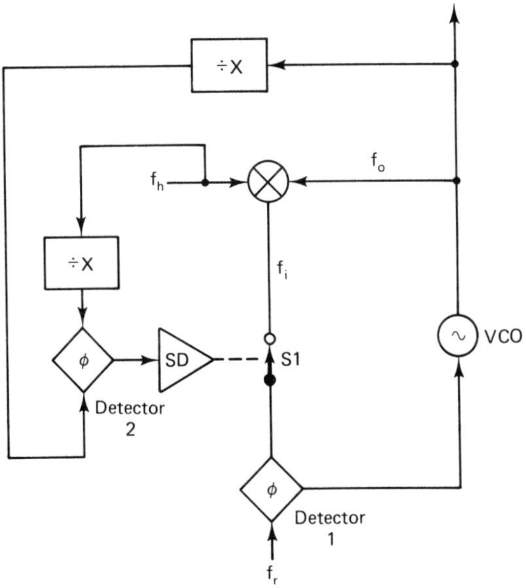

Figure 9-4. Additional circuits to maintain correct loop tracking

If, during a transient, output f_o is less than f_h, two conditions will affect normal loop operation. With $f_o < f_h$ and $f_i < f_r$, the output of phase detector 1 will cause f_o to increase. As f_o increases, f_i decreases to zero frequency at $f_o \fallingdotseq f_h$ and then increases until $f_i \approx f_r$ and normal lock is obtained. If $f_o << f_h$ so that $f_i > f_r$, the output of phase detector 2 will continue to decrease f_o, and phase lock in the tracking loop will never occur. This condition is prevented by making phase detector 2 open switch S1 when $f_h/X > f_o/X$; thus $f_r > f_i$ since $f_i = 0$ and phase detector 1 causes VCO frequency f_o to increase. When $f_o \geq f_h$, phase detector 2 closes the loop, and normal lock can be obtained when $f_o \approx f_h$.

The frequency dividers of fixed value X were added to this example in the event that the values of f_o and f_h were too high for phase-frequency devices to operate properly. If f_o and f_h are at lower values where these devices can operate correctly, the X dividers would not be required.

REFERENCES

1. Connell, P., and Hull, B., "Synthesized Source Joins Analog Purity, Digital Speed," *Electronic Design* (September 30, 1982), pp. 109–16.
2. Fadrhous, J., "Blend a Wideband PLL," *Electronic Design* (July 5, 1976), pp. 46–51.

10

Wideband Microwave Frequency Synthesizers

Starting with this chapter and continuing in the remaining chapters describing frequency synthesizer techniques, the emphasis is on block diagrams and examples that illustrate a variety of designs that have been used for broadband microwave frequency synthesizers. The equations detailed in Chapters 7, 8, and 9 can be used to predict the power spectral density of the multiple loops contained in these designs.

One of the more common technical approaches using YIG devices [1] is illustrated in Fig. 10-1. Beginning at the upper left, a high-frequency reference (f_h) drives an impulse generator, a step-recovery diode circuit, that outputs many harmonics well into the microwave bands. The YIG-tuned filter is controlled with a programmable direct current for selection of a desired nth harmonic of f_h. The selected harmonic at a frequency nf_h is mixed with a sample of the output frequency f_o prior to division by 10. This fixed divider is usually needed to lower the IF in order to use phase-frequency detectors since these digital devices have a low upper-frequency limit.

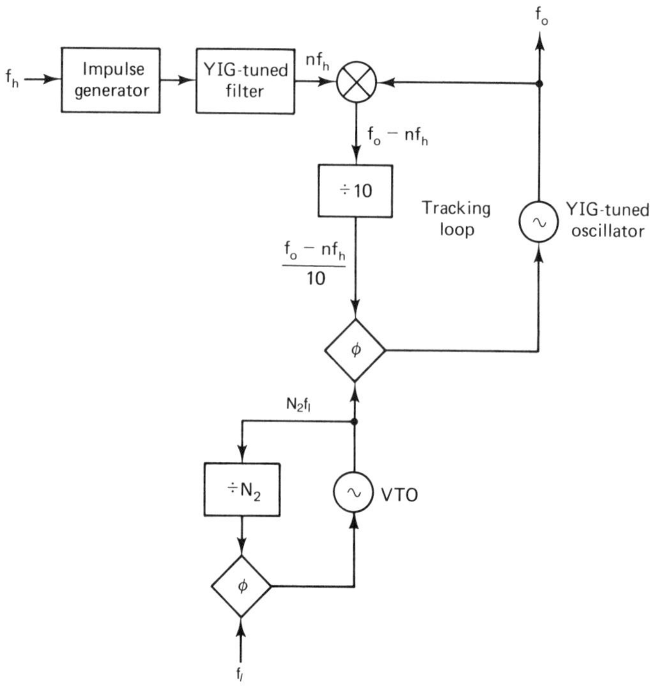

Figure 10-1. Frequency synthesis with YIG devices

At the phase-frequency detector inputs, when phase-lock has been achieved, the two frequencies are equal:

$$\frac{f_o - nf_h}{10} = N_2 f_l \tag{10-1}$$

and the YIG-tuned oscillator's frequency is given by

$$f_o = 10\,N_2 f_l + nf_h \tag{10-2}$$

letting $nf_h = N_i f_h$, where N_i is a specific harmonic of f_h:

$$f_o = 10\,N_2 f_l + N_i f_h \tag{10-3}$$

As an example of how this technique is employed, assume that we want a microwave synthesizer tunable in 100-kHz increments from 4 to 5.9999 GHz. Letting the programmable divider loop have a reference f_l of 10 kHz and letting N_2 be selectable from 2000 to 2999 in unity increments makes the value of 10 $N_2 f_l$ programmable from 200 to 299.9 MHz and supplies all 10-MHz, 1-MHz, and 100-kHz tuning steps. Our impulse generator with an input of 100 MHz is designed to maximize those harmonics in the 3800- to 5700-MHz band with 100-MHz spacing. The YIG-tuned filter is assumed to be programmable in frequency with D/A converters supplying the correct direct current for positioning the YIG's bandpass filter center on the N_i harmonic; thus, as we tune across the band from 3800 to 5700 MHz in steps of 100 MHz, the programmable divider loop cycles from $N_2 = 2000$ to $N_2 = 2999$ within each 100-MHz step.

This type of synthesizer design can provide a signal output (frequency f_o) with excellent spectral purity since a YIG-tuned oscillator is used in the tracking loop and the loaded Q of these oscillators is very high; however, this design approach has two major limitations. One is that frequency set-on time is slow since two YIG devices are employed, and these devices are very slow when compared to varactor-tuned devices. Second, positioning of the YIG-tuned filter's center frequency on selectable harmonics can be very complex and costly. In these current-controlled filters, highly stable components are required to keep the desired harmonic within the filter band. Small amounts of drift associated with the direct current from the YIG driver, or drifts within the YIG, can require heaters for stable temperature operation; therefore, periodic calibration of the YIG's frequency-versus-current relationship may be required due to component aging. In the following discussions, an alternate design approach developed by the author is explained. The technique was put into production, and hundreds of these specialized synthesizers have been in use over the years. A YIG-tuned oscillator is not required but can be used when tuning speed is not important.

LOOPS WITH VARIABLE PHASE-DETECTOR REFERENCE FREQUENCY

Let us examine two techniques in which the phase-frequency detector's reference frequency is varied, in a defined manner, to cause any selected harmonic to be programmable in defined discrete increments. The first technique is illustrated in Fig. 10-2. An electronically tuned oscillator (ETO) with an output frequency f_o is phase-locked to reference frequencies f_r and f_s. The programmable divider values of $N + S$ and N depend on the particular value of the nth harmonic value desired at f'_o, the harmonic generator output frequency. For this technique to work properly, the value of N in the dividers must equal the selected numerical value of the nth harmonic.

When the loop is phase-locked, the phase-detector inputs are equal:

$$\frac{f_o - f_r}{N + S} = \frac{f_s}{N} \tag{10-4}$$

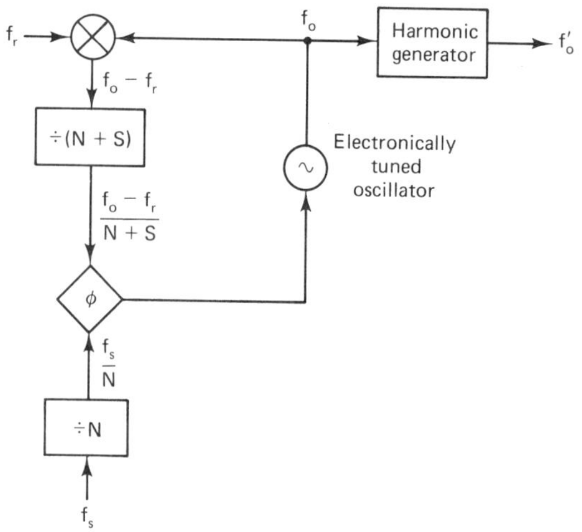

Figure 10-2. PLL with variable reference frequency and down converter

and f_o can be expressed as

$$f_o = \frac{(N + S)}{N} f_s + f_r \tag{10-5}$$

With $f'_o = Nf_o$, Eq. (10-5) is expanded to

$$f'_o = N(f_r + f_s) + Sf_s \tag{10-6}$$

The importance of Eq. (10-6) is that variables N and S with constants f_r and f_s can make f'_o tunable in increments of f_s at any desired harmonic N_i!

DESIGN EXAMPLE OF VARIABLE REFERENCE FREQUENCY

As an example of how this technique is employed, let us assume that we will allow all harmonics of an impulse generator, the harmonic generator in Fig. 10-2, to be present; however, we want to make any selected harmonic in a band of frequencies extending from 2000 to 4000 MHz to vary in 10-MHz increments. Our design would require the following:

1. f_r = 90 MHz
2. f_s = 10 MHz
3. N, variable from 20 to 40
4. S, variable from 0 to 9

and our equation is given by

$$f'_o = (N \times 100\text{ MHz}) + (S \times 10\text{ MHz}) \tag{10-7}$$

For the 20th harmonic, $N = 20$ and

$$f'_o = 2000\text{ MHz} + (S \times 10\text{ MHz}) \tag{10-8}$$

As S is varied from 0, 1, 2, . . ., 8, 9, the discrete values of f'_o are

2000 MHz

2010 MHz

2020 MHz

2030 MHz

2040 MHz

2050 MHz

2060 MHz

2070 MHz

2080 MHz

2090 MHz

At the 21st harmonic ($N = 21$), starting at 2100 MHz, the processes are repeated and so on at all remaining harmonics. Table 10-1 is an abbreviated list of frequencies showing the dependence of f'_o on all other parameters of this example. The complete table could be put in a look-up table in a memory to automatically set the frequency dividers to the correct value for all output frequencies ranging from 2000 to 4000 MHz.

We noted earlier that when all harmonics from the impulse generator are present, we can select a particular harmonic with a tracking loop or a tunable filter. We will discuss these methods later. However, the major application for this technique is to provide a variable microwave reference in a microwave frequency synthesizer. Before proceeding to the second variable reference technique, there are a few more items worth discussing.

We can calculate the tuning range required of the VCO using Eq. (10-5):

$$f_o = \frac{N + S}{N} f_s + f_r \tag{10-5}$$

The minimum value occurs when $S = 0$, and for our example it is equal to 100 MHz. The maximum value of f_o occurs when $(N + S)/N$ is maximum

TABLE 10-1

f'_o (MHz)	nth Harmonic	$\div(N + S)$	$\div N$	S
2000	20	20	20	0
2010	20	21	20	1
2020	20	22	20	2
2090	20	29	20	9
2100	21	21	21	0
2110	21	22	21	1
2190	21	30	21	9
2200	22	22	22	0
2210	22	23	22	1
2290	22	31	22	9
2900	29	29	29	0
2910	29	30	29	1
2990	29	38	29	9
3500	35	35	35	0
3510	35	36	35	1
3590	35	44	35	9
3800	38	38	38	0
3810	38	39	38	1
3890	38	48	38	9
3900	39	39	39	0
3910	39	40	39	1
3990	39	48	39	9
4000	40	40	40	0

and is equal to 29 ÷ 20; thus the maximum value of f_o is

$$\left(\frac{29}{20} \times 10\right) + 90 = 104.5 \text{ MHz}$$

The worst-case phase noise on the output f'_o will occur when the quantity $N + S$ is a maximum. Within the loop bandwidth, the phase noise present on the phase reference at f_s/N is multiplied by $N + S$ at f_o and is multiplied again by the nth harmonic of f'_o. Thus the SNR at f'_o is degraded by 20 log $(N + S)N$, from the SNR of phase noise at reference frequency f_s/N.

Loop stability can be maintained by divider $N + S$ in the loop, but the minimum and maximum values of $N + S$ need to be included for the total range of this divider. Loop bandwidth, which determines tuning speed, should not be allowed to be greater than 1/20 of the minimum value of f_s/N (the lowest value of loop reference frequency) in order to maintain adequate phase margin.

SECOND TYPE OF VARIABLE REFERENCE FREQUENCY

The second technique involving a variable frequency as a reference to the phase-frequency detector is shown in Fig. 10-3. The main advantage of this design over that shown in Fig. 10-2 is that reference source f_r and the mixer are not required. The major disadvantage is that the feedback divider is much larger; this has the potential of degrading the output signal's spectral purity.

The design required to use this approach is a two-step process, as follows: First, do an analytical design of the technique described for Fig. 10-2 to compute the required values of f_r, f_s, N, and S. The second step is to use these values in our model of Fig. 10-3. For these conditions, we equate the two-phase detector inputs:

$$\frac{f_o}{\dfrac{f_r + f_s}{f_s} N + S} = \frac{f_s}{N} \tag{10-9}$$

which, at the ETO output, is

$$f_o = \left[\frac{f_s}{N}\left(\frac{f_r + f_s}{f_s}\right) N + S\right] \tag{10-10}$$

and in terms of the nth harmonic at f'_o

$$f'_o = N(f_r + f_s) + Sf_s \tag{10-11}$$

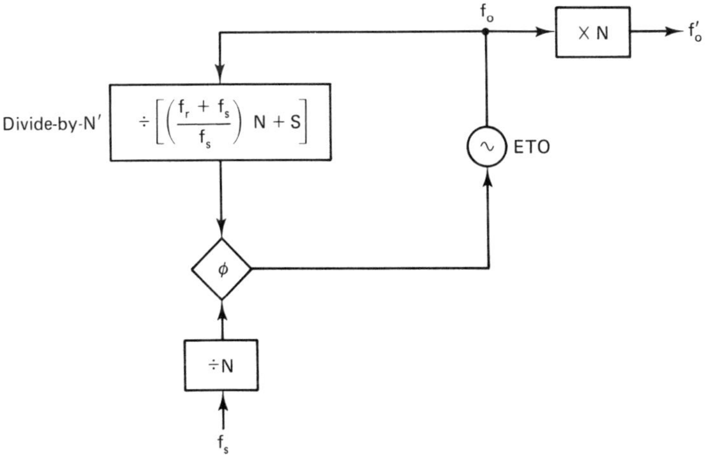

Figure 10-3. PLL with variable reference frequency

Equation (10-11) is identical to Eq. (10-6); therefore, the output control of frequency is identical. In the actual fabrication of this technique, only a single reference frequency at f_s is required, as we noted previously, and divider N' is much larger. The application of this second approach with variable references to the phase detector is for microwave synthesizers in which larger tuning increments are used at the higher microwave bands. In Chapter 11 are presented several design examples that use these two techniques for synthesizers in the microwave frequency bands.

ESTIMATE OF PHASE-NOISE SIDEBANDS

To estimate the phase-noise sidebands of these types of synthesizers, we need to know the input reference frequency ratio of single-sideband to carrier power ratio for the example illustrated in Fig. 10-4. This example is for the synthesizer model of Fig. 10-2 with the addition of a low-noise amplifier (LNA) that has a 2-dB noise figure. The analysis is for the condition where N and $N + S$ are simultaneously at a maximum from

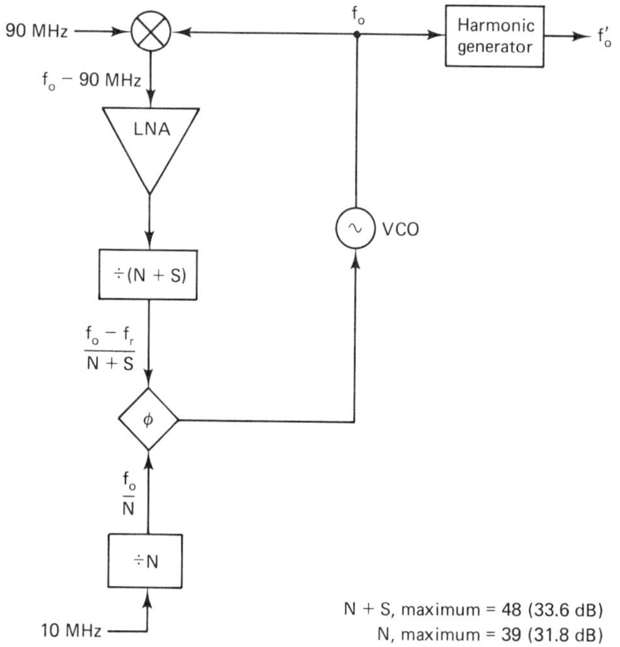

Figure 10-4. Example for estimating phase noise sidebands

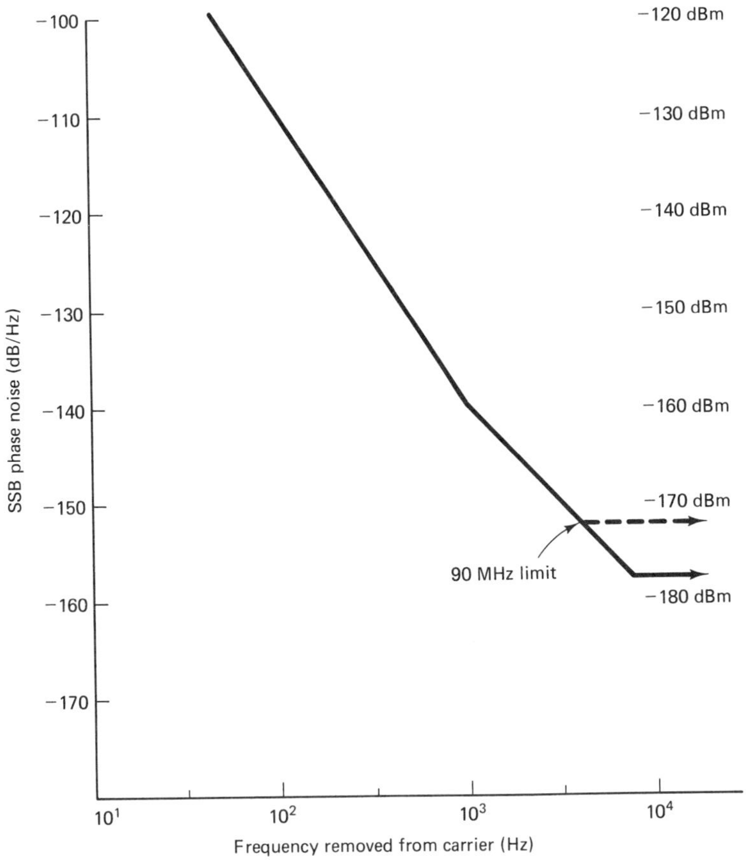

Figure 10-5. 90 MHz reference SSB phase noise

Table 10-1. These divider values are used in Eq. (10-5) to estimate the phase noise of output f_o for the worst case when dividers are at a maximum in the loop. Estimates of SSB phase noise assumed for the VCO, 90- and 10-MHz references, and divider output limits are typical of devices currently used in lower-frequency phase-locked loop synthesizers in the following figures.

In Fig. 10-5 the phase-noise ratio expressed in decibels per hertz is shown graphically for a 90-MHz VCXO, assumed to be phase-locked to a single low-frequency synthesizer reference. The output of the VCXO is at a power level of 0 dBm; therefore, the vertical scale at the left can also represent the phase-noise power in dB below 0 dBm. For example, the phase noise at a frequency removed (f_m) of 1 kHz is −140 dBm. In order to have low spurious signal generation in the mixer, this level is attenuated

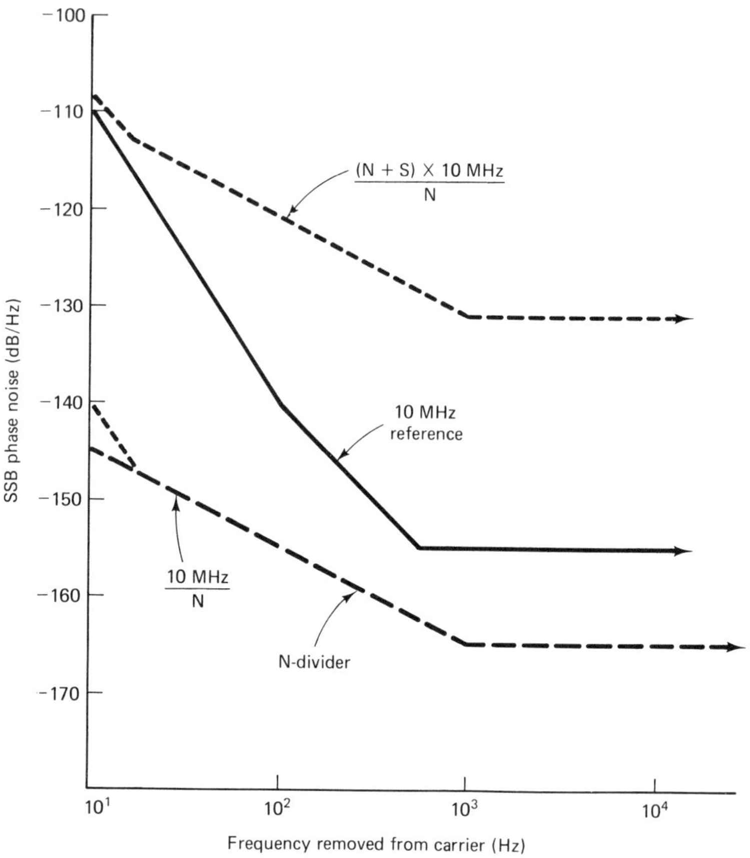

Figure 10-6. 10 MHz reference SSB phase noise

by 13 dB for application of −13 dBm to the mixer. A typical mixer, with 7 dB of conversion loss, will cause the mixer output of (f_o − 90 MHz) to be reduced to −20 dBm.

The scale at the right in Fig. 10-5 is the SSB phase-noise power expressed in dBm for the mixer IF output at −20 dBm; thus at f_m of 1 kHz, the SSB phase noise is at −160 dBm. With an amplifier noise figure of 2 dB, kTB is increased from −174 dBm (1 Hz, BW) to −172 dBm; this power level establishes the lower limit to the 90-MHz source, referenced to the mixer output. For this reason, the ratio of SSB phase noise to carrier power is reduced to the dashed-line, 90-MHz limit.

From Eq. (10-5),

$$f_o = f_{90} + (N + S)f_{10}/N$$

where f_{90} is the 90-MHz reference and f_{10} is the 10-MHz reference, and if the open-loop gain is very large, the VCO phase noise can approach the limits established by the 90-MHz reference in Fig. 10.5.

In Fig. 10-6 the SSB phase noise of the 10-MHz reference is shown as the solid line for a 10-MHz crystal oscillator reference. This reference is divided by N and is reduced by 10 log N; for $N = 39$, it is reduced by 31.8 dB to the longer dashed line of the divider's output SSB noise floor. Since the 31.8-dB reduction is to a level well below the divider's limit, this divider sets a limit to SSB phase noise with respect to the 10-MHz reference except for a low value of f_m near 10 Hz, as illustrated with the short dashed line. This is the condition where the 31.8-dB SSB phase noise is slightly larger than the divider SSB phase-noise threshold. The divided output, for Eq. (10-5), is multiplied by $N + S$ and increases the phase noise by

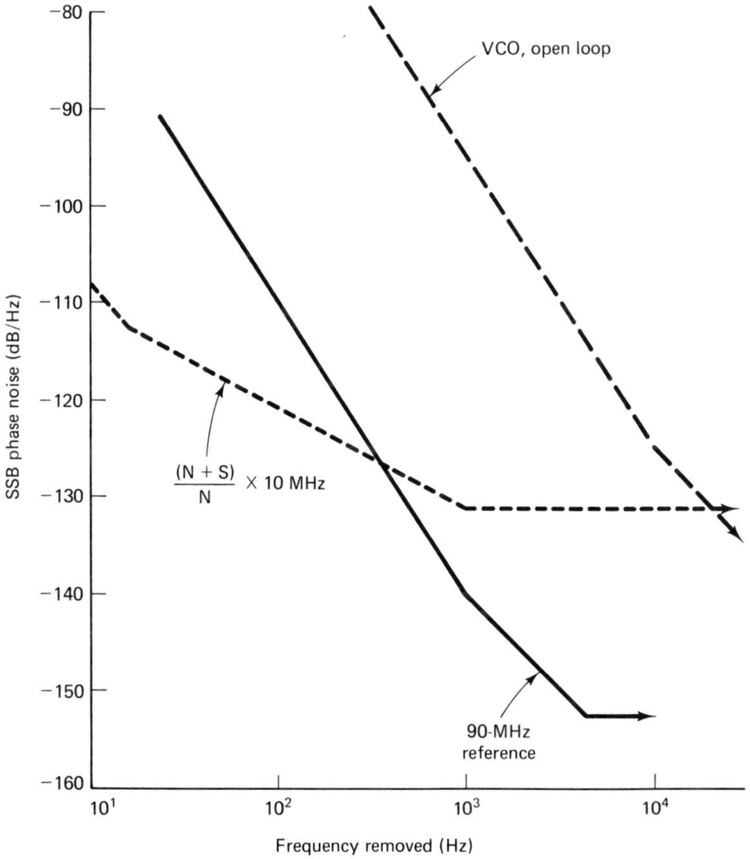

Figure 10-7. Output f_o SSB phase noise

33.6 dB to the level shown with the short dashed lines at the top of the figure.

In Fig. 10-7 the composite SSB phase noise of the 90-MHz reference and the divided or multiplied 10-MHz reference are shown as the solid and short dashed lines. In addition, the open-loop VCO-SSB phase noise is shown as the longer dashed lines. When the open-loop gain is very large, the VCO-SSB phase noise can be lowered to the levels established by the two input references; however, it is limited to the higher of the two values. For example, at f_m lower than 100 Hz, the 90-MHz phase noise is larger and establishes the locked VCO sideband noise spectrum. At f_m between 100 and 1000 Hz, the divided and multiplied 10 MHz becomes the limit as f_m is increased greater than 1000 Hz.

In Chapter 9 an approximate limit to open-loop gain was stated as 1/20 of the reference to the phase detector. In this example, 1/20 of f_{10}/N is approximately 12.8 kHz; therefore, large open-loop gain cannot be obtained at $f_m = 10$ kHz. The VCO phase noise will be reduced by $20 \log |1 + A/N|$, and when this value exceeds the difference between open-loop VCO-SSB phase noise and composite reference SSB phase noise, the net closed-loop SSB phase noise will be equal to the combined reference limits. As f_m is increased, the closed-loop SSB phase-noise ratio will be reduced, and when f_m is equal to the value of f_m where A/N is less than 1, no reductions in VCO-SSB phase noise can be obtained. To obtain the lowest levels of VCO phase noise, the optimum value of f_m for $A/N = 1$ is in the vicinity of equality of open-loop VCO phase noise and reference phase noise. In this example, these two SSB phase-noise ratios are equal at f_m slightly larger than 10 kHz.

REFERENCE

1. Harp, J. C., *The WJ-1250 Frequency Synthesizer System.* Watkins-Johnson Application Note 100072-076A, 1976.

11

Microwave Frequency Synthesizer Design Examples

2- TO 4-GHz FREQUENCY SYNTHESIZER

In this chapter are presented three separate design examples in which the variable phase-detector reference technique is used. These three designs are very similar to microwave frequency synthesizers designed by the author and put into operation in synthesized signal generators using each of these techniques. In reference [1], a 2-to-6.2 GHz synthesizer design is described in which a variable phase-detector reference is used and is noted as an M/N loop. The first example, illustrated in Fig. 11-1, uses the configuration of Fig. 10-2 as the harmonic loop at the upper left in Fig. 11-1. With f_r equal to 240 MHz and f_s equal to 10 MHz, the output at f_o is given by

$$f_o = \frac{N + S}{N} \times 10 \text{ MHz} + 240 \text{ MHz} \tag{11-1}$$

With $f'_o = Nf_o$

$$f'_o = (N \times 250 \text{ MHz}) + (S \times 10 \text{ MHz}) \tag{11-2}$$

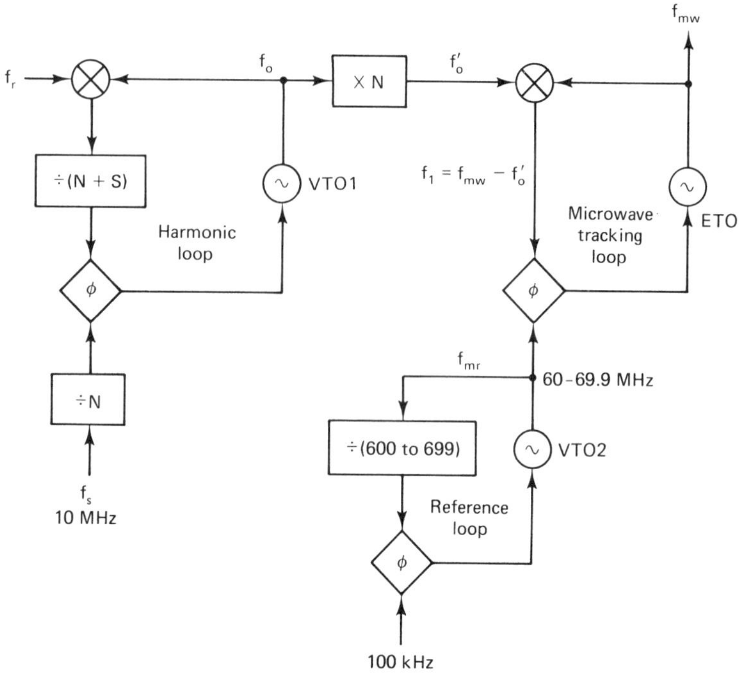

Figure 11-1. Microwave frequency synthesizer with variable reference

and makes the microwave reference f'_o tunable in 10-MHz increments at each nth harmonic of f_o.

The other reference frequency (f_{mr}) to the microwave loop is derived from a single-loop synthesizer, referred to as the reference loop, with output from VTO2 tunable in 100-kHz increments from 60 to 69.9 MHz. The selection of the numerical value of this reference frequency is critical to making the synthesizer operate correctly and is defined as follows:

> The maximum value of f_1 including the initial mistuning error of the open-loop ETO source frequency must not exceed the sum of $(f_r + f_s) \div 2$.

This requirement is derived from receiver image frequency considerations where an interfering signal is located at a frequency equal to the sum or difference of the selected frequency and twice the value of the receiver IF. In our synthesizer design, our lowest separations between harmonics of f_o occur when step S is equal to 0. Under this condition, the separation between harmonics is 250 MHz, and at an IF of 125 MHz the ETO frequency is midway between two harmonics and may not allow the microwave tracking loop to lock onto the correct harmonic of f_o.

Making the reference frequency f_{mr} approximately equal to one-fourth the minimum harmonic separation of 250 MHz allows an open-loop tuning error of the ETO to be nearly at a maximum value. In this example, with f_{mr} maximum $\approx$ 70 MHz, the allowable tuning error is

$$(125 - 70)\ \text{MHz} = 55\ \text{MHz} \tag{11-3}$$

To avoid a condition where f_1 could go to 0, the lowest value of f_{mr} at 60 MHz is used in the following relation:

$$(60 - 60)\ \text{MHz} = 0\ \text{MHz} \tag{11-4}$$

Thus our worst-case limits of tuning error are +55 MHz and −60 MHz; in a practical design, the limits should be reduced to ±50 MHz to avoid potential locking problems at the allowable error limits.

The output frequency from the microwave ETO, f_{mw}, is defined by

$$f_{mw} = f'_o + f_{mr} \tag{11-5}$$

and when combined with Eq. (11-2), the result is

$$f_{mw} = (N \times 250\ \text{MHz}) + (S \times 10\ \text{MHz}) + (60\ \text{to}\ 69.9)\ \text{MHz} \tag{11-6}$$

In Table 11-1, several values of the output f_{mw} are listed as a function of the variables, f'_o, N, S, and f_{mr}. In this design for a synthesizer that can

TABLE 11-1

f_{mw}	f'_o	N	S	f_{mr}
2000 to 2009.9	1940	7	19	60 to 69.9
2010 to 2019.9	1950	7	20	60 to 69.9
2020 to 2029.9	1960	7	21	60 to 69.9
2030 to 2039.9	1970	7	22	60 to 69.9
2040 to 2049.9	1980	7	23	60 to 69.9
2050 to 2059.9	1990	7	24	60 to 69.9
2060 to 2069.9	2000	8	0	60 to 69.9
2070 to 2079.9	2010	8	1	60 to 69.9
2300 to 2379.9	2240	8	24	60 to 69.9
2310 to 2319.9	2250	9	0	60 to 69.9
2550 to 2559.9	2490	9	24	60 to 69.9
2560 to 2569.9	2500	10	0	60 to 69.9
2800 to 2809.9	2740	10	24	60 to 69.9
3060 to 3069.9	3000	12	0	60 to 69.9
3300 to 3309.9	3240	12	24	60 to 69.9
3560 to 3569.9	3500	14	0	60 to 69.9
3800 to 3809.9	3740	14	24	60 to 69.9
3810 to 3819.9	3750	15	0	60 to 69.9
3900 to 3909.9	3840	15	9	60 to 69.9
3990 to 3999.9	3930	15	18	60 to 69.9
4000	3940	15	19	60 to 69.9

be tuned from 2000 to 4000 MHz in 100-kHz tuning increments, the minimum value of N is 7 and the maximum value is 15. These limits of N are needed to compute the minimum and maximum value of $N + S$ to make the harmonic loop stable with adequate phase margin.

8- TO 18-GHz FREQUENCY SYNTHESIZER

In this section, a design for a broadband microwave frequency synthesizer operating from 8 to 18 GHz is derived and is based on using the technique of Fig. 10-3 for the harmonic loop. The synthesizer block diagram is illustrated in Fig. 11-2. The synthesizer output frequency f_{mw} is given by

$$f_{mw} = f'_o + 10\, f_{mr} \tag{11-7}$$

with reference frequency f_{mr} tunable in 10-kHz increments from 20 to 29.99

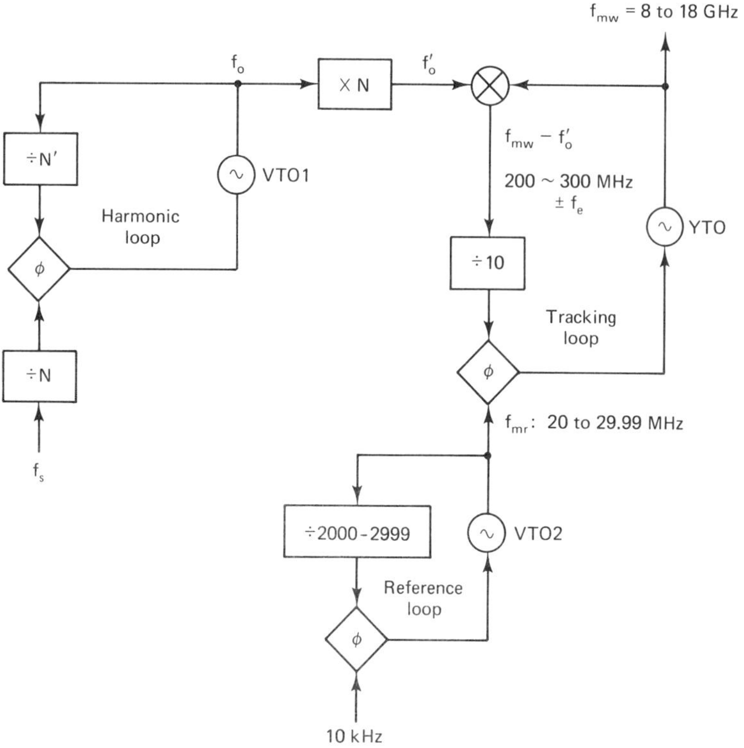

Figure 11-2. Wideband microwave frequency synthesizer, 8 to 18 GHz

MHz. This reference's contribution in output f_{mw} is to provide the tuning increments of 100 kHz, 1 MHz, and 10 MHz.

The tracking loop contains a YIG-tuned oscillator that is tunable from 8 to 18 GHz by coarse-tuning circuits (excluded from Fig. 11-2) that initially set the YTO to within 150 MHz of the desired synthesized frequency. The maximum tuning error f_e of ± 150 MHz is one of the main advantages of this design and allows the use of low-accuracy tuning circuits and helps to lower cost. This large allowable initial error is obtained by designing the harmonic loop VTO1 frequency to have a minimum value of 1 GHz.

With the tracking loop IF equal to $200 \sim 300$ MHz $\pm f_e$, where f_e is the YTO tuning error, the minimum IF is $200 - 150$ MHz and the maximum IF is $300 + 150$ MHz; therefore, in the worst case, the IF can vary from 50 to 450 MHz prior to phase-lock in this loop. Once phase-lock is obtained, the IF will vary from 200 to 299.9 MHz.

As stated earlier, the harmonic loop is designed to have a minimum output frequency f_o of 1 GHz and will make any nth harmonic tunable in 100-MHz increments. To make this design achieve these objectives requires following the two-step procedure described for Fig. 10-3: First model the design, as was shown in Fig. 10-2, with the following values:

$$f_r = 900 \text{ MHz}; \quad f_s = 100 \text{ MHz}$$

with the values of N and S to be defined later. The second step is to convert this model to the configuration shown in Fig. 10-3 for determining the value of divider N'.

The value of S will be variable from 0 to 9 since each harmonic of f_o at f'_o will be varied over a 900-MHz band in 100-MHz increments. The value of N is determined as follows:

$$f'_o = f_{mw} - 10\, f_{mr} = N(f_r + f_s) + S(f_s) \tag{11-8}$$

$$N = \frac{f_{mw} - 10\, f_{mr} - S(f_s)}{f_r + f_s} \tag{11-9}$$

Since each unique value of N remains constant as S varies from 0 to 9, and $10\, f_{mr}$ varies from 200 to 299.9 MHz, let $S = 0$ and $10\, f_{mr} = 200$ MHz in Eq. (11-9) and it simplifies to

$$N = \frac{f_{mw} - 200 \text{ MHz}}{1000 \text{ MHz}} \tag{11-10}$$

Since N must be an integer, the frequencies f_{mw} for computing the minimum and maximum values of N are

$$f_{mw} = 7200 \text{ MHz}; \quad N = 7$$

$$f_{mw} = 17{,}200 \text{ MHz}; \quad N = 17$$

At the lowest output frequency of 8000 MHz, f'_o = 7800 MHz with N = 7 and S = 8.

Recalling that the value of divider N' is given by:

$$N' = \left(\frac{f_r + f_s}{f_s}\right) N + S \tag{11-11}$$

the minimum value of N' is

$$N' \text{ (minimum)} = (10 \times 7) + 0 = 70 \tag{11-12}$$

and the maximum value of N' is

$$N' \text{ (maximum)} = (10 \times 17) + 9 = 179 \tag{11-13}$$

We are now in a position to make a table relating N, S, and N' to specific frequencies of the synthesized output f_{mw}. Table 11-2 is a list of some of these values, and all remaining ones can be readily computed. We begin with the lowest values, $N = 7$ and $S = 8$, as previously computed. The value of N' is, of course, computed from Eq. (11-11). All synthesized values from 8000 to 8099.9 MHz are controlled by f_{mr}; simply stated,

TABLE 11-2

Line	f_{mw}	f'_o	N	S	N'
1	8,000	7,800	7	8	78
2	8,100	7,900	7	9	79
3	8,200	8,000	8	0	80
4	8,300	8,100	8	1	81
5	8,400	8,200	8	2	82
6	8,500	8,300	8	3	83
7	8,600	8,400	8	4	84
8	8,700	8,500	8	5	85
9	8,800	8,600	8	6	86
10	8,900	8,700	8	7	87
11	9,000	8,800	8	8	88
12	9,100	8,900	8	9	89
13	9,200	9,000	9	0	90
14	10,100	9,900	9	9	99
15	10,200	10,000	10	0	100
16	11,100	10,900	10	9	109
17	11,200	11,000	11	0	110
18	12,100	11,900	11	9	119
19	13,200	13,000	13	0	130
20	14,100	13,900	13	9	139
21	14,200	14,000	14	0	140
22	15,100	14,900	14	9	149
23	17,200	17,000	17	0	170
24	18,000	17,800	17	8	178

frequency reference f_{mr} controls tuning increments smaller than 100 MHz. For the convenience of keeping our computations simple, we maintain f_{mr} at 20 MHz while computing the values of N' from N and S.

In line 3, the value of f'_o when divided by 1000 MHz is an integer; therefore, $S = 0$, and N is determined by inspection when the three zeros are removed. This process is repeated at lines 5, 7, 9, 11, 13, 15, 17, 19, 21, and 23. The value of S can also be determined by inspection as shown in lines 2, 4, 6, 8, 10, 12, 14, 16, 18, 20, 22, and 24. The value of S is the hundreds value of f'_o. The next step is to take the sum of N and S to compute the divider numerical value N' of Eq. (11-11) for Fig. 11-2.

The final step in our synthesizer design would be to design a digital processor to compute these values automatically as a function of external programming for selection of the microwave output f_{mw}.

PROGRAMMABLE MULTIPLIERS

In the third design example, a simplified version of this technique applied to a microwave synthesizer designed by the author is presented. To begin, we take the position that a single-loop synthesizer as illustrated in Fig. 11-3(a) is equivalent to a programmable multiplier as illustrated in Fig. 11-3(b). To extend the technique of using a phase detector with a variable reference to synthesizers with single loops, we consider the single loop to be the harmonic generator in the previous examples.

In Fig. 11-4 a single-loop microwave frequency synthesizer contains a fixed divide-by-10 prescaler and a programmable divider M. The output of the single loop f_{mw} is equal to the sum of reference f_{mr} and 10 M times

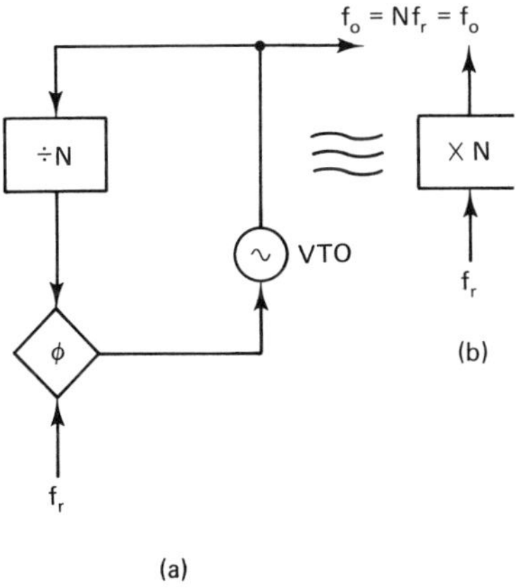

Figure 11-3. Single loop synthesizer as a programmable multiplier

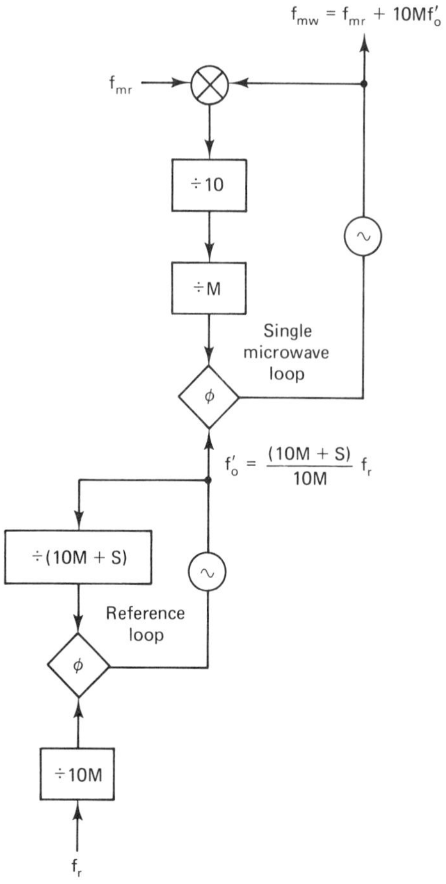

Figure 11-4. Single-loop microwave frequency synthesizer, variable reference

reference f'_o. In the reference loop and in the single loop, the following equations determine the microwave output f_{mr}:

$$f_{mw} = f_{mr} + 10Mf'_o \tag{11-14}$$

$$f'_o = \frac{10M + S}{10M} f_r \tag{11-15}$$

Combining Eqs. (11-14) and (11-15),

$$f_{mw} = f_{mr} + 10Mf_r + Sf_r \tag{11-16}$$

If we let microwave reference f_{mr} be tunable in 100-MHz increments, and with f_r, a 1-MHz reference, the divider M need only vary from 10 to 19 to cause the output f_{mw} to step in 10-MHz increments. The value of S in the reference loop controls the 1-MHz increments, and S will vary from 0 to 9. The main advantage of this third example is that frequency reference f'_o is never less than 1 MHz; therefore, the single-loop bandwidth can be ten times larger than if f'_o were a constant at 0.1 MHz.

Let us now examine two graphs that illustrate the synthesis techniques discussed for Figs. 11-1 and 11-2. In Fig. 11-5 the multiple mixer IF responses are shown as a function of the synthesizer output frequency, specifically for when the output is tuned from 3310 to 3319.1 MHz as the microwave loop's reference is varied from 60 to 69.9 MHz. The solid diagonal lines represent the multiple harmonics present at the mixer input and the mixer's IF outputs resulting from these harmonics when $f_{mw} > f'_o$. The dashed diagonal lines are IF image responses when $f'_o > f_{mw}$ and are potential problem responses.

As reference f_{mr} is stepped from 60 to 69.9 MHz, the output tracks from 3310 to 3319.9 MHz as shown by the intersection of the short dashed vertical lines of f_{mw} intersecting with the horizontal solid lines representing

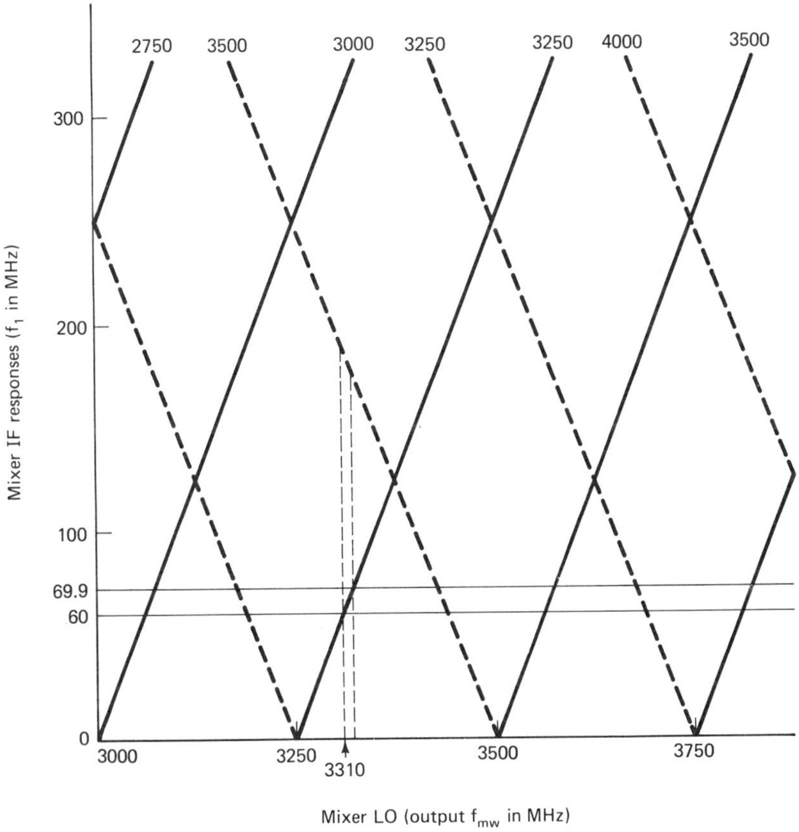

Figure 11-5. Graph of mixer inputs in the 2- to 4-GHz synthesizer

f_1. From Fig. 11-5, there exists an image response present between the IF values of 150 and 200 MHz caused by the input harmonic of 3500 MHz. However, this image response is much greater than the 60- to 69.9-MHz IF and does not present any errors (after being attenuated by a low-pass filter) caused by locking to the wrong harmonic.

When f_{mw} is initially tuned to 3310 MHz with an error of ±50 MHz prior to phase lock, the IF can range from 10 to 110 MHz. At 110 MHz, the desired mixer output from the 3250-MHz harmonic approaches the undesired output caused by the image response of the 3500-MHz harmonic. If the initial set-on error of f_{mw} causes f_{mw} to equal 3375 MHz, the two responses are equal at 125 MHz, and interference of these two signals can cause the microwave loop not to lock to the correct harmonic; therefore,

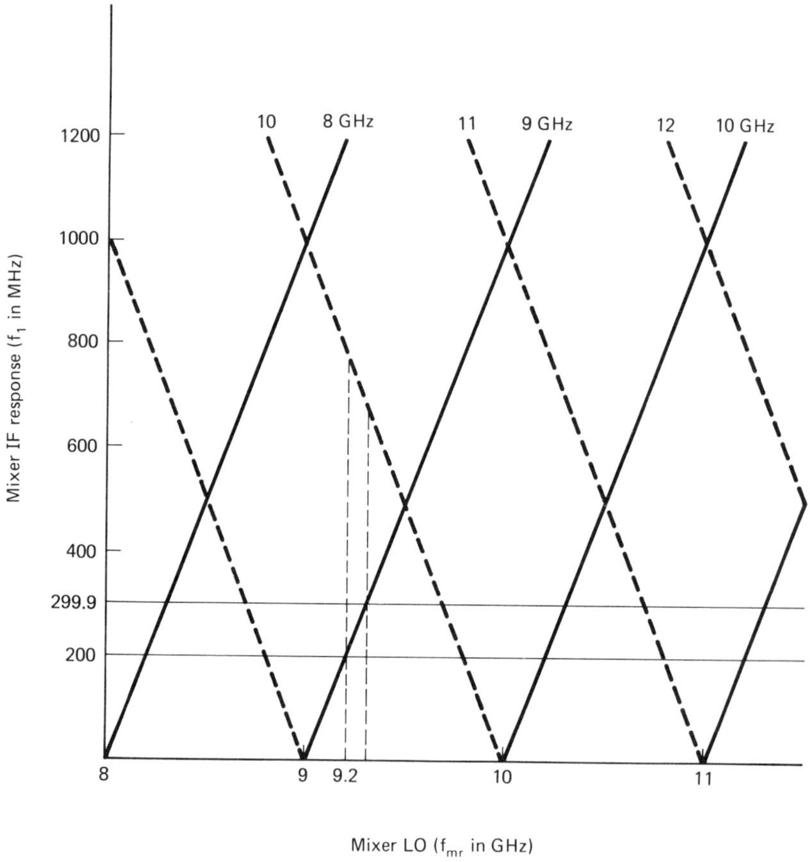

Figure 11-6. Graph of mixer inputs in the 8- to 18-GHz synthesizer

it is important to limit the initial mistuning error of f_{mw} to less than ± 50 MHz.

If the loop is designed to increase the frequency of f_{mw} when $f_{mr} > f_1$, mistuning errors of f_{mw} can be lower than 3250 MHz but should not be low enough to cause the 3250-MHz image response to exceed f_{mr} as derived from Fig. 11-5 for f_{mr} less than 3250 MHz. Thus a reasonable range of initial set-on errors can extend from -100 to $+50$ MHz.

The example of Fig. 11-2 for an 8- to 18-GHz synthesizer has the advantage of allowing larger initial set-on errors, and this is important with frequencies as large as 18 GHz.

In Fig. 11-6 the same type of mixer IF responses for multiple inputs are shown for the 8- to 18-GHz synthesizer. When the output is 9200 to 9399.9 MHz obtained from the 9-GHz input and the YTO frequency of f_{mw}, the mixer IF outputs are 200 to 299.9 MHz and greater than 600 MHz. If the initial set-on frequency tuning errors are less than 150 MHz, the mixer output will be within 50 to 450 MHz for any programmed value of f_{mr}, and false lock will not occur.

When the synthesizer is designed for the phase detector to increase the value of f_{mw} when the IF, divided by 10, is lower than $f_{mr}/10$, the lower limit of mistuning error can be increased considerably more than 150 MHz.

From Fig. 11-6, the initial value of 9.2 GHz can be lower than 9 GHz by as much as 200 MHz before false locks can occur; therefore, a reasonable limit of allowable error is -400 and $+150$ MHz, allowing for a low-cost course-tuning circuit for the 8- to 18-GHz YTO.

REFERENCE

1. Astrof, K. L., "Frequency Synthesis in a Microwave Signal Generator," *Hewlett-Packard Journal* (November 1977), pp. 8–15.

12

Novel Design Techniques

In this chapter, we examine three novel design techniques, two of which use a modification of a PLL with a variable phase-detector reference frequency. In Fig. 12-1 the position of "divide-by-N" and "divide-by-(N + S)" are interchanged from the configurations discussed in Chapters 10 and 11 to derive synthesized time increments rather than synthesized frequency. Analysis follows:

SYNTHESIZED TIME INCREMENTS

At the phase detector,

$$\frac{f_o}{N} = \frac{f_r}{N + S} \tag{12-1}$$

Letting τ_o represent the period of f_o and τ_r represent the period of f_r,

$$\frac{1}{N\tau_o} = \frac{1}{(N + S)\tau_r} \tag{12-2}$$

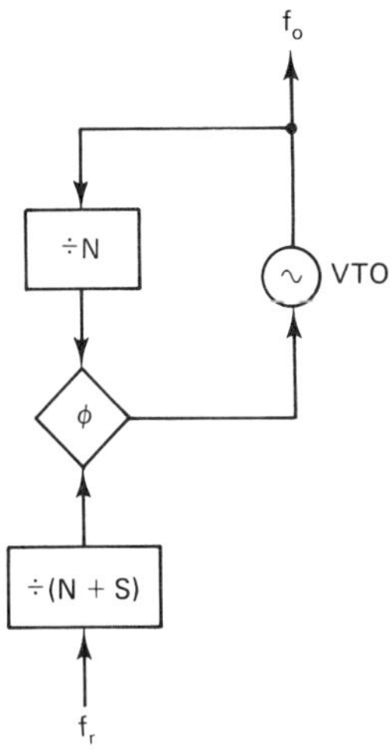

Figure 12-1. Synthesized increments of time

$$\tau_o = \left(\frac{N + S}{N}\right)\tau_r = \left(1 + \frac{S}{N}\right)\tau_r \qquad (12\text{-}3)$$

One of the uses of Eq. (12-3) is to make the output frequencies time-base-programmable in discrete percentage increments. For example, assume that a variable period ranging from 1 to 2 ns is needed in increments of 0.01 ns. We would make the reference frequency f_r 1 GHz to make τ_r in Eq. (12-3) equal to 1 ns. Next we let divider N be fixed at 100 and let divider $N + S$ be programmable from 100 to 200 in steps of 1 by increasing S from 0 to 100. Substituting these values into Eq. (12-3),

$$\tau_o = \frac{100 + S}{100}\text{ ns} = \left(1 + \frac{S}{100}\right)\text{ ns} \qquad (12\text{-}4)$$

As S is increased from 0 to 1, 2, 3, etc., time base τ_o starts at 1 ns and increases in steps equal to 0.01 ns. If greater resolution were required, 0.001 ns, for example, we would make $N = 1000$ and vary S from 0 to 1000. In these examples, the tuning range of our VTO is 2:1 as computed from Eq. (12-1) in the form of

$$f_o = \left(\frac{N}{N + S}\right) f_r \qquad (12\text{-}5)$$

To make time base τ_o start at the larger value of 2 ns, we make S begin at its largest value and decrease it to 0 to make τ_o decrease. The main advantage of the circuit illustrated in Fig. 12-1 is that very small timing intervals can be generated and selected in small, discrete increments with a simple single-loop synthesizer.

FRACTIONAL FREQUENCY SYNTHESIS

Occasionally, a constant frequency source is required to be derived from a reference standard where the constant frequency source's accuracy is not too important and values in the range of 10 to 100 ppm are acceptable. For example, in music an octave contains 12 tones with adjacent tones being multiples of $2^{1/12}$. One can use a computer to approximate the fractional value of

$$2^{1/12} \approx 1.059463 \tag{12-6}$$

One fractional value that is within 45 ppm of this value is

$$1 + \frac{6}{101} = \frac{107}{101} \approx 1.059406 \tag{12-7}$$

From inspection of the ratio 107/101, we can make the single loop of Fig. 12-2 cause f_o to be equal to $(107 \div 101)\, f_r$; therefore, f_o is approximately equal to $2^{1/12}$ times f_r.

Table 12-1 lists f_o divider values and f_r divider values obtained by an iteration procedure to approximate $2^{n/12}$ to within an accuracy of a few percent for simulation of an octave on a piano keyboard.

In this example, an output slightly larger than the standard reference was needed; when the required output is smaller than the standard reference, the divider values are interchanged, and we would have in our example

$$f_o = \frac{f_r}{2^{1/12}} \tag{12-8}$$

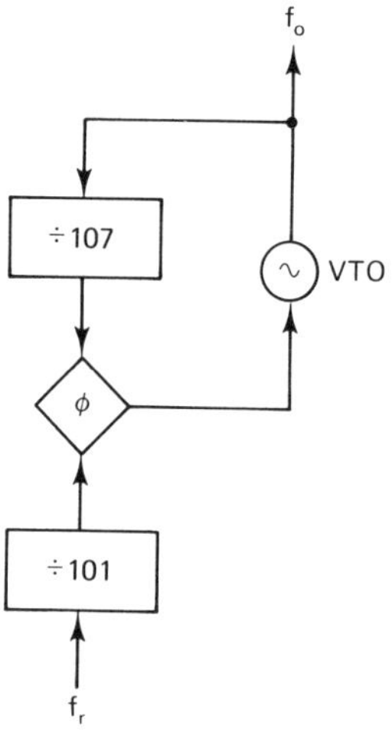

Figure 12-2. PLL to approximate an increase of f_r by $2^{1/12}$

TABLE 12-1

Multiplier	f_o Divider	f_r Divider
$2^{1/12}$	107	101
$2^{2/12}$	229	204
$2^{3/12}$	69	58
$2^{4/12}$	126	100
$2^{5/12}$	199	149
$2^{6/12}$	140	99
$2^{7/12}$	298	199
$2^{8/12}$	316	199
$2^{9/12}$	673	400
$2^{10/12}$	713	400
$2^{11/12}$	755	400

Note: This form of fractional frequency synthesis is not the same as described in reference [1].

FAST-TUNED REFERENCE FREQUENCY GENERATOR

In some applications, a microwave-synthesized reference signal is required where larger tuning increments (25 to 250 MHz, for example) are adequate. This requirement would exist in a fast-tuned microwave frequency synthesizer similar to Fig. 11-1, with the harmonic loop replaced with the block diagram shown in Fig. 12-3.

In Fig. 12-3 reference f_r is assumed to be a highly stable frequency source operating at a fixed frequency, usually in the 50- to 500-MHz frequency range. Harmonic generator N has an output containing many harmonics of f_r where any discrete value of N is represented by N_i. The IF from the mixer output must be less than $f_r/2$ when the microwave VTO is coarse-tuned by the D/A converter to a frequency near the final phase-locked frequency value.

At phase lock, the IF frequency is equal to the phase-detector reference frequency:

$$f_{mw} \pm N_i f_r = f_r/4 \tag{12-9}$$

and this reduces to

$$f_{mw} = (N_i \pm 1/4) f_r \tag{12-10}$$

and according to Eq. (12-10), the output frequency f_{mw} can be phase-locked

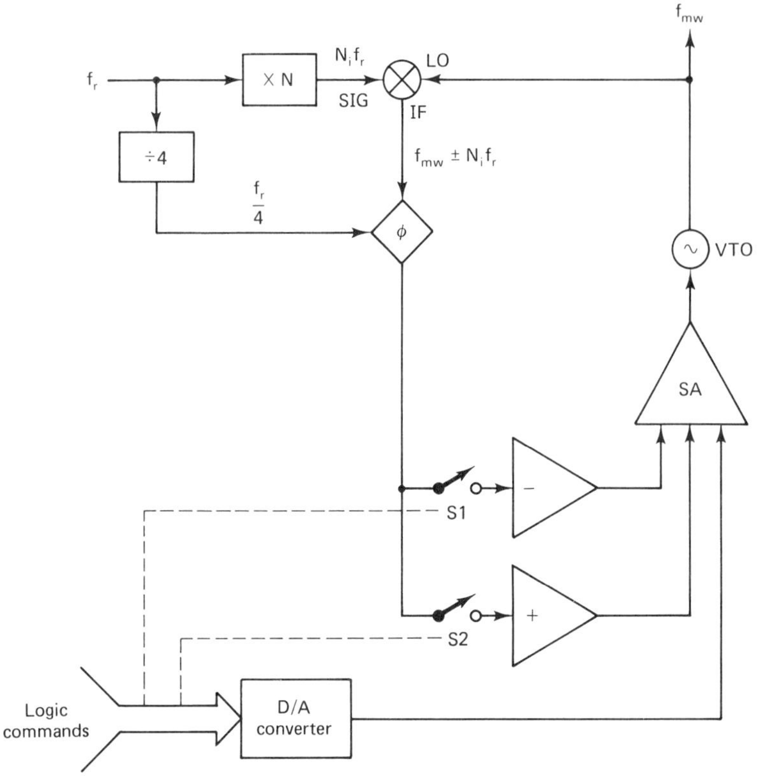

Figure 12-3. Fast tuned synthesizer with large tuning increments

at

$$(N_i - 1/4)f_r$$

$$(N_i + 1/4)f_r$$

$$(N_i + 1 - 1/4)f_r$$

$$(N_i + 1 + 1/4)f_r$$

$$(N_i + 2 - 1/4)f_r$$

$$(N_i + 2 + 1/4)f_r$$

.

.

.

$$(N_i + k - 1/4)f_r$$

To make all of the preceding frequencies available for phase lock at the selected value, it is necessary for the open-loop VTO frequency to be coarse-tuned, by the D/A converter, to less than $\pm 1/4\, f_r$ of the required phase-locked value. The function of switches S1 and S2 combined with the inverting and noninverting amplifiers is to make the loop correction drive the VTO to phase lock and not from it. This condition can occur because the loop is to be locked with the LO port frequency greater than and less than the signal (SIG) port frequency.

For example, with $f_r = 200$ MHz, we have listed the outputs of f_{mw} as a function of N_i and the proper state of the switches (see Table 12-2). One advantage of this technique is that from a 200-MHz spectral-line generator, stable phase lock is obtained at frequencies spaced 100 MHz apart. There are other significant advantages as well:

1. The RF power level of the locked VTO can be considerably greater than the harmonics of f_r. This results in a significant cost savings since the oscillator requires only one stage of gain, whereas an amplifier-filter combination would require many gain states.
2. With the phase-detector reference frequency at a high megahertz value, the loop bandwidth can be in the low megahertz range; fast tuning is obtained, and a relatively noisy oscillator's spectrum can be made equal to approximately N_i times the very clean reference spectrum of f_r.
3. Using a harmonic loop of either Fig. 10-2 or Fig. 10-3 as the reference frequency f_r makes the output tunable in increments considerably smaller than $f_r/2$.

The synthesis design example is shown graphically in Fig. 12-4. The desired phase-lock IF frequency of 50 MHz occurs when $|f_{mw} - N_i f_r| =$

TABLE 12-2

f_{mw} (MHz)	N_i	S1	S2
3950	20	open	closed
4050	20	closed	open
4150	21	open	closed
4250	21	closed	open
4350	22	open	closed
4450	22	closed	open
4550	23	open	closed
4650	23	closed	open
4750	24	open	closed
4850	24	closed	open
4950	25	open	closed

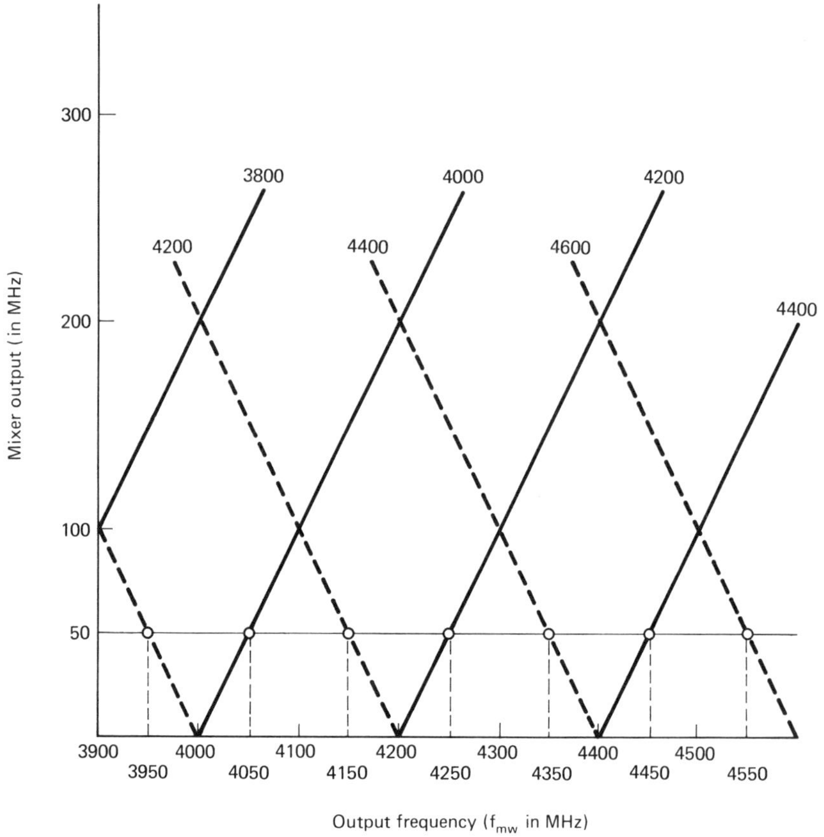

Figure 12-4. Graph of mixer IF output responses

50 MHz; these lock frequencies are shown as small circles on the 50-MHz IF response. As in the other examples similar to this technique, the initial mistuning error needs to be bounded in order to prevent false locks. For example, at f_{mw} = 4250 MHz, the error must not exceed +50 or −100 MHz. At f_{mw} = 4150 MHz, the error must not exceed −50 or +100 MHz.

REFERENCE

1. Goodall, D., "Synthesized Signal Generator Breaks Cost Performance Barrier," *Microwaves and RF* (October 1984), pp. 153–56.

13

Measuring Indirect-Type Synthesizer Performance

The techniques used to measure phase-lock loop bandwidth and frequency switching speed of the output signal are described in this chapter. Techniques are described to measure the magnitude of $1 \div (1 + A)$ where A is the open-loop gain of the single-loop synthesizer illustrated in Fig. 7-2 in Chapter 7 [1]. This figure is repeated as Fig. 13-1 for the reader's convenience. For this loop,

$$|A| = \frac{GH(S)K_v K_\Phi}{f_m} \tag{13-1}$$

where f_m is the modulating frequency and G, $H(S)$, K_v, and K_Φ are the loop component's transfer functions. Recalling Eq. (7-1) in Chapter 7 and repeating it here as Eq. (13-2):

$$f_o = \frac{K_v}{1 + A} V_{no} + \frac{H(S)K_v}{1 + A} V_{na} + \frac{A}{1 + A} f_r \tag{13-2}$$

We can indirectly measure the magnitude of $1 \div (1 + A)$ by opening the loop and injecting a small variable-frequency voltage at the oscillator's input designated as V_{no}.

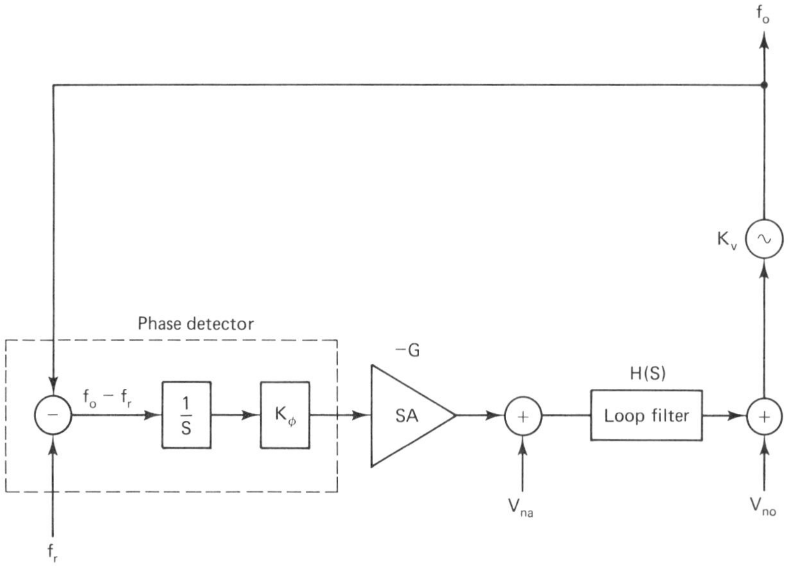

Figure 13-1. Performance of the single-loop synthesizer

Let the injected signal be described as

$$V_s = V_p \sin (2\pi f_m t) \tag{13-3}$$

where the peak voltage V_p is adjustable and the frequency f_m can also be varied. With the loop opened, the output spectrum of f_o is observed on a spectrum analyzer. Starting with $V_p = 0$, the amplitude V_p is increased slowly until the sidebands on f_o are in a range of 20 to 30 dB down and removed from f_o by a value equal to the modulating frequency f_m. The loop is then closed, and the sideband amplitude in terms of "dB down" is remeasured. The difference between measured "open-loop" and "closed-loop" values is a measure of the magnitude of $1 \div (1 + A)$.

Sometimes it is easier to begin the measuring processes with values of f_m considerably greater than loop bandwidth and then decrease frequency f_m. Since the "dB down" value is given by

$$\text{dB down} = 20 \log \frac{2f_m}{\Delta f} \tag{13-4}$$

where the peak deviation Δf is equal to $K_v \times V_p$ of Eq. (13-3), it is necessary to make adjustments to V_p each time f_m is changed to maintain a constant value of "dB down" in Eq. (13-4).

Following these processes for each value of f_m (first open-loop measurements and then closed-loop measurements) to compute the magnitude of $1 \div (1 + A)$, we then calculate the magnitude of the open-loop function $1 + A$. For a stable loop, the magnitude of $1 + A$ should not be less than 0.5, since this value would cause 6-dB peaking in the loop. A lower limit of $|1 + A|$ is suggested to be 0.7 in order to have no more than 3 dB of loop peaking.

Closed-loop gain is most easily measured by using a signal generator capable of being frequency-modulated for the reference frequency f_r. A spectrum analyzer is needed to set the sidebands in the range of 20 to 30 dB down on f_r at the modulation frequency f_m [2]. Beginning with values of f_m considerably less than unity closed-loop gain, the sidebands on f_o will be equal to the "dB down" value of f_r because we are observing the function $A \div (1 + A)$ in Eq. (13-2) and this function (closed-loop gain) is equal to unity for large values of open-loop gain A. As modulation frequency f_m is increased, the magnitude of $|A|$ decreases, and in the vicinity of $|A| = 1$ we will observe the largest amount of closed-loop gain peaking if the gain margin is too small. To maintain adequate gain and phase margin in the loop, the closed-loop gain peak should never exceed 6 dB; a maximum limit of 3 dB is recommended.

LOOPS WITH PROGRAMMABLE DIVIDERS

Figure 13-2 is Fig. 8-2 of Chapter 8, repeated here for the reader's convenience. The equation describing the signal's output frequency f_o is

$$f_o = \frac{K_v}{1 + A/N} V_{no} + \frac{H(S)K_v}{1 + A/N} V_{na} + \frac{A}{1 + A/N} f_r \qquad (13\text{-}5)$$

This is Eq. (8-1) from Chapter 8.

Examination of Eq. (13-5) reveals that we can obtain information on the magnitude of $1 \div (1 + A/N)$ by injecting a variable-amplitude signal generator and letting it represent V_{no} as in the previous discussion. Now instead of measuring the magnitude of $1 \div (1 + A)$, we will measure the magnitude of $1 \div (1 + A/N)$. Letting A' equal A/N makes the measurement of A' identical to the previous discussion.

Once we have determined the approximate value of modulation frequency f_m where the magnitude of $(1 + A/N)$ is a minimum, we can use this information to measure the closed-loop gain. The closed-loop gain for reference f_r is given by $A \div (1 + A/N)$. For values of modulation frequency f_m considerably less than the value of f_m to cause $1 + A/N$ to be a minimum, the closed-loop gain is equal to the divider value N. As the modulating frequency f_m is increased in variable reference f_r, the closed-loop gain may go to a peak value prior to rolling off. For values of f_m

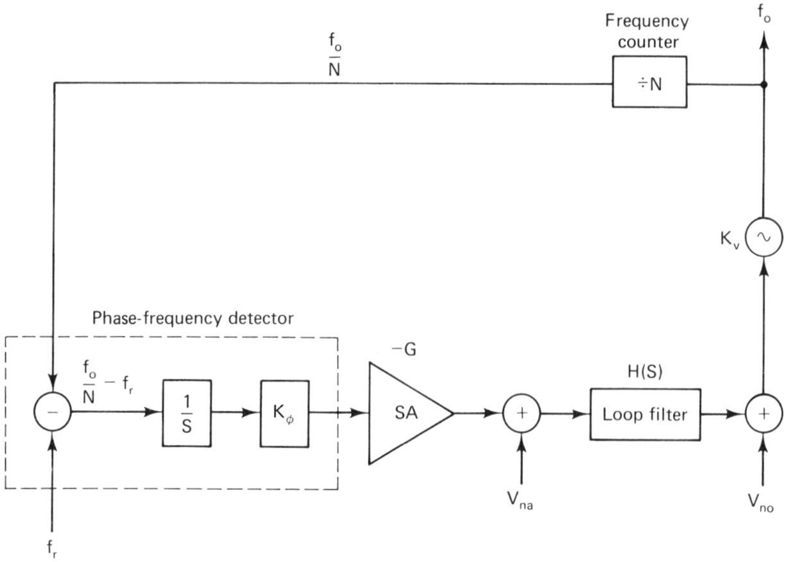

Figure 13-2. Loop model containing a programmable counter

that cause $A/N << 1$, the gain is equal to $|A|$ and is significantly less than N, its low-frequency gain value.

MEASURING FREQUENCY SET-ON TIME

The switching speed, frequency set-on time, of indirect synthesizers can be measured in two separate locations in the synthesizer. The first location to be discussed is at the phase-frequency detector; however, this particular measurement requires the addition of a phase detector placed in parallel with the phase-frequency detector as illustrated in Fig. 13-3. The reason for adding the phase detector is that this device has two states. When the loop is out of lock, the phase detector output is a beat frequency and is equal to the frequency difference between the two inputs and is clearly observed on the oscilloscope as an out-of-lock condition. The measurement setup diagrammed in Fig. 13-3 is similar to Fig. 5-5 in Chapter 5, where all of the measurement equipment is synchronized to a single clock frequency, the loop reference f_r.

For the phase-frequency detector in an out-of-lock condition, the output is a constant voltage level. As the loop goes into a phase-locked state, the voltage transitions smoothly to a final value, and it is difficult to define exactly when lock occurred. However, the phase detector output

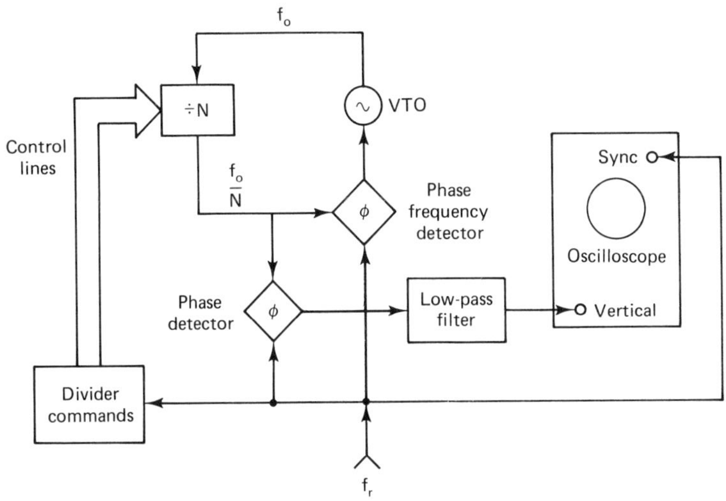

Figure 13-3. Test to measure switching speed

will change abruptly from a beat frequency to a transient change in voltage level as the loop locks and the phase differences between the two inputs transition smoothly to a final steady-state value.

The single problem associated with the measurement described previously is that the output at f_o is N times greater than the phase transients measured at the phase detector. Since frequency is the derivative of phase with respect to time, the large phase transients occurring in f_o result in a transient error frequency in f_o. The transient error can be computed by multiplying the phase detector–measured values by the value of divider N.

To make a measurement directly on the output f_o requires a measurement fixture as illustrated in Fig. 13-4. In this fixture, the synthesizer under test reference frequency standard f_r is used as the clock to synchronize all of the equipment. A second frequency synthesizer is required to provide a fixed frequency LO to the double-balance mixer as shown. The digital controller's function is to cause the synthesizer under test to switch alternately to two frequencies, one of which is equal to the LO frequency and the other to any fixed value. Under these conditions, the observed waveform will contain a beat frequency when the synthesizer frequency is not equal to the LO frequency and numerous transients when locking to a frequency equal to the LO frequency. The large variety of observed transients is caused by locking at many values of phase differences at initial lock prior to settling at the final steady-state value.

In the event that two frequency synthesizers are not available, the LO frequency and the DBM are deleted from Fig. 13-4. A discriminator

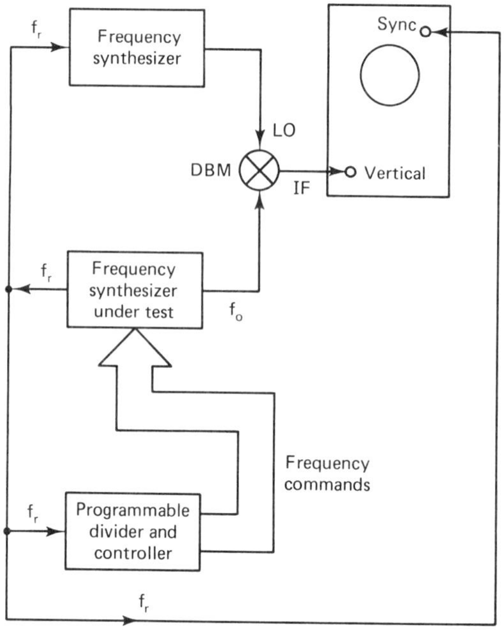

Figure 13-4. Direct measurement of the output switching speed

would be placed at the output of the synthesizer under test, and discriminator responses would be measured on the scope. To calibrate the discriminator, steady-state output voltage levels from the discriminator are marked on the scope face as a function of several different synthesizer frequencies. After the scope has been calibrated to the discriminator, the synthesizer is then repeatedly switched to two different frequencies, and the transitions between these two values can be observed on the scope. The major disadvantage of this measurement is that the out-of-lock and transient-after-lock states cannot be measured separately; in addition, the measurement accuracy is limited by resolution limits of making measurements on a CRT.

DETAILED SWITCHING TIME MEASUREMENTS

The details of making switching time measurements of indirect-type synthesizers are examined in Fig. 13-5. The measurement being performed is similar to the technique discussed in Fig. 13-3. However, the method of controlling the synthesized output and observing frequency switching transients is also similar to the measurement technique of Fig. 13-4.

The example of Fig. 13-5 is for the following conditions:

1. The output frequency f_o is programmable in 1-MHz steps from 400 to 799 MHz.
2. Frequency divider N is controlled by three sets of BCD logic lines for increments of hundreds, tens, and units.
3. The control line driver is capable of driving any combination of lines up to a maximum of 12.
4. The VCO phase-frequency detector has a 1-MHz reference.

Since the loop's reference is 1 MHz, we assume the loop BW to be approximately 50 kHz; therefore, the control lines are switched at a much

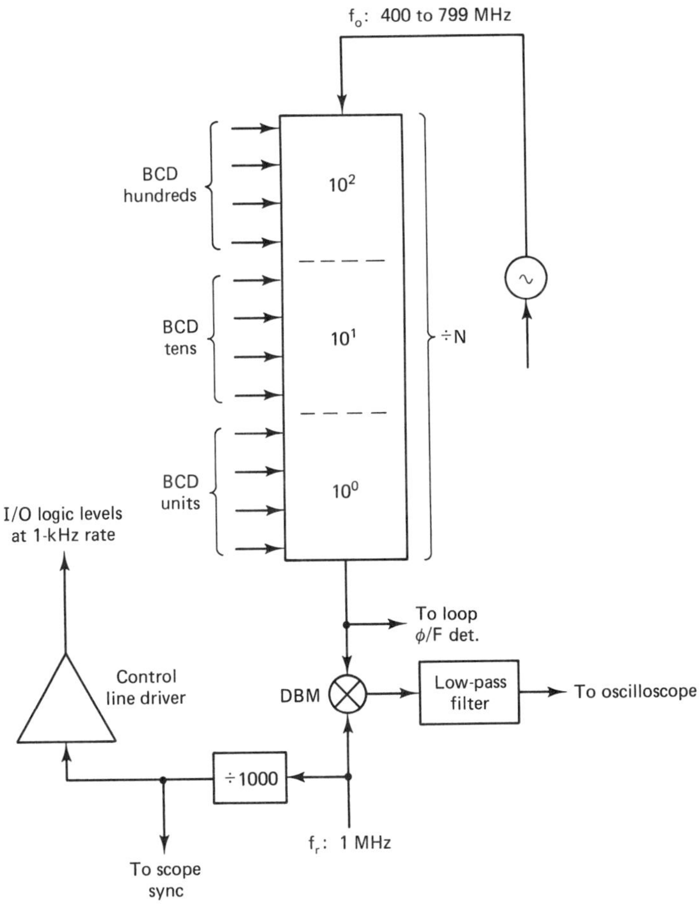

Figure 13-5. Details of measurement technique applied to Figure 13-3

lower rate of 1 kHz to allow ample time to observe the transients as the programmed lines are switched to logic 1 and 0 levels.

For the purpose of observing these transients, a double-balanced mixer is used as the measurement device. Since harmonics of the DBM inputs are present at its output, a low-pass filter is added to reduce their levels to very small values that do not obscure the measured transients during and after phase lock of the VCO. Initially, let us assume that we want to measure the loop's acquisition for small steps in the vicinity of 400 MHz and our programmable divider is controlled by logic high levels for turning on a BCD bit and logic low levels for turning off a BCD bit.

The divider would be programmed for the four hundreds bit by keeping only this control line high. All other BCD control lines are at a low except for the units control line. This line is controlled by the control line driver and is switched at a 1-kHz rate between high and low levels; therefore, the synthesizer's output frequency will alternate between 400 and 401 MHz. Starting with the VCO in the phase-locked state at 400 MHz, the VCO may initially break lock when the divider programming is changed from 400 to 401, and it may assume a new value of $400 \pm \Delta f$ MHz. This represents an offset of $(400 \pm \Delta f)/401$ and causes the DBM to have an output beat frequency equal to $(1 \mp \Delta f)/401$. The condition of measuring these transient conditions is controlled by selecting either positive or negative edge synchronization on the oscilloscope to observe 400 MHz switched to 401 MHz or 401 MHz switched to 400 MHz.

With output frequency steps as small as 1 MHz, it is possible that the loop may not break lock because of the large amount of division in the loop. Larger-frequency step size is obtained by letting the control line driver switch larger bit values of the divider's programming lines. If the frequency step size is too large, switching from 799 to 400 MHz, the initial beat frequency out of the phase detector is approximately 399/400 MHz and may not be initially observed on the scope due to the low-pass filter's 3-dB cutoff frequency. For these reasons, the step size should be selected carefully to assure that initial beat frequencies can be observed as they proceed to 0 frequency as the loop relocks.

REFERENCES

1. Hall, J. A., "Applications of a Phase-modulated Signal Generator," *Hewlett-Packard Journal* (July 1975), pp. 21–24.
2. Rands, R., "Phase-Modulation Techniques Simplify Analysis of Phase-locked Loops," *Electronics* (August 21, 1975), pp. 94–98.

14

Combining Direct and Indirect Techniques

Frequency synthesizer designs are generally neither 100% direct nor 100% indirect. In many cases, the design approach contains a mixture of both techniques to achieve a design optimized for performance at the lowest hardware cost. In this chapter and in Chapters 15 and 16, synthesizer designs in which both techniques are combined are examined.

We begin with a double-loop microwave synthesizer where low phase noise and small size were the major factors for selecting the synthesizer design approach; in addition, measured sideband noise data are presented. In Chapter 15 more complex designs are examined with emphasis on transforming a direct-type design into an indirect-type design. Chapter 16 presents a design approach where large-bandwidth phase-locked loops combined with fast-tuned, directly derived references achieve high-speed frequency agility in the microwave region.

The double-loop microwave synthesizer was designed to be tunable in 10-MHz increments. Frequency stability of a few ppm was adequate; therefore, a 125-MHz crystal oscillator was selected for the synthesizer's standard reference, as shown at the left side in Fig. 14-1. The wideband loop for the 2-GHz VTO was needed to reduce sideband noise levels from

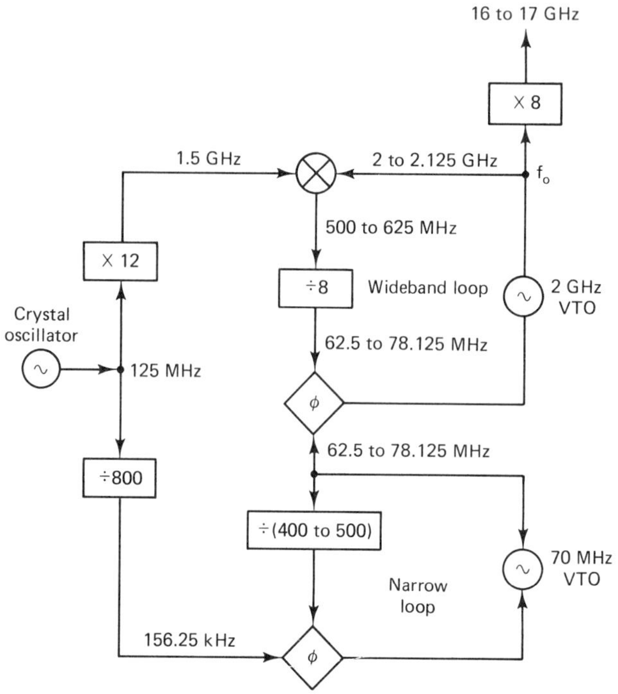

Figure 14-1. Combined direct-indirect synthesizer

close to the carrier to 100 kHz removed from the carrier; therefore, a loop bandwidth exceeding 300 kHz was required.

From the single-reference standard, two references are derived using simple direct techniques of multiplication by 12 and division by 800. The frequencies at 1.5 GHz and 0.156250 MHz are used as references in the two loops.

The low-frequency loop, output from 62.5 to 78.125 MHz, has a narrow loop bandwidth in order to minimize the effects of programmable divider noise at the 70-MHz VTO output. The programmable divider (400 to 500) combined with the phase detector's 0.15625-MHz reference provides a variable reference to the microwave loop; this reference is tunable in 0.15625-MHz increments from 62.5 to 78.125 MHz.

At the 2-GHz VTO output, the divide-by-8 within the loop combined with the 1.5-GHz reference defines the output f_o as

$$f_o = 1500 \text{ MHz} + 8(62.5 \text{ to } 78.125 \text{ MHz}) \tag{14-1}$$

$$f_o = 2000 \text{ to } 2125 \text{ MHz} \tag{14-2}$$

and f_o is tunable in 1.25-MHz increments. After multiplying f_o by 8, the Ku-band output from 16 to 17 GHz is tunable in 10-MHz increments.

To summarize the design approach:

1. The 125-MHz crystal oscillator provides a high-spectral-purity signal, after multiplication by 12, to the wideband loop. The sideband noise levels in this reference are not dominant in the phase-locked 2-GHz VTO's output close to the carrier because the 1.5-GHz signal was derived from a crystal oscillator with very low sideband phase noise.
2. Division by 800 in the low-frequency reference makes the sideband noise spectrum of the 0.15625-MHz reference to be mainly dependent on the divider's noise floor and not the 125-MHz reference's noise sidebands.
3. The low-frequency VTO's spectral components are the major contributor to the microwave VTO's sideband noise up to 40 kHz removed from the carrier.

Since the low-frequency, 70-MHz VTO loop is narrowband, the programmable divider's effect on the microwave VTO's spectrum is dominant from close in to a few kilohertz removed from the carrier.

Measurements of the 2- to 2.125-GHz VTO sideband noise power were made with a microwave discriminator and converted to a carrier-to-noise ratio [1,2]. The data are listed in Table 14-1.

In theory, the closed-loop SNR should be equal to the sum of open-loop SNR and open-loop gain; in this design at low values of frequency removed (f_m), the gain is very large and the synthesizer's SNR was close

TABLE 14-1

f_m (kHz)*	Open-Loop Oscillator SNR (dBc/Hz)	Closed-Loop Oscillator SNR (dBc/Hz)	Open-Loop Gain (dB)
5	70	100	43
10	80	107	33
20	89	112	23
40	98	118	16
80	106	120	9
160	114	120	3
320	121	118	0
640	128	126	−2
1280	135	135	

*Frequency removed from the carrier's average frequency. Measurement error for all data, 2 to 3 dB.

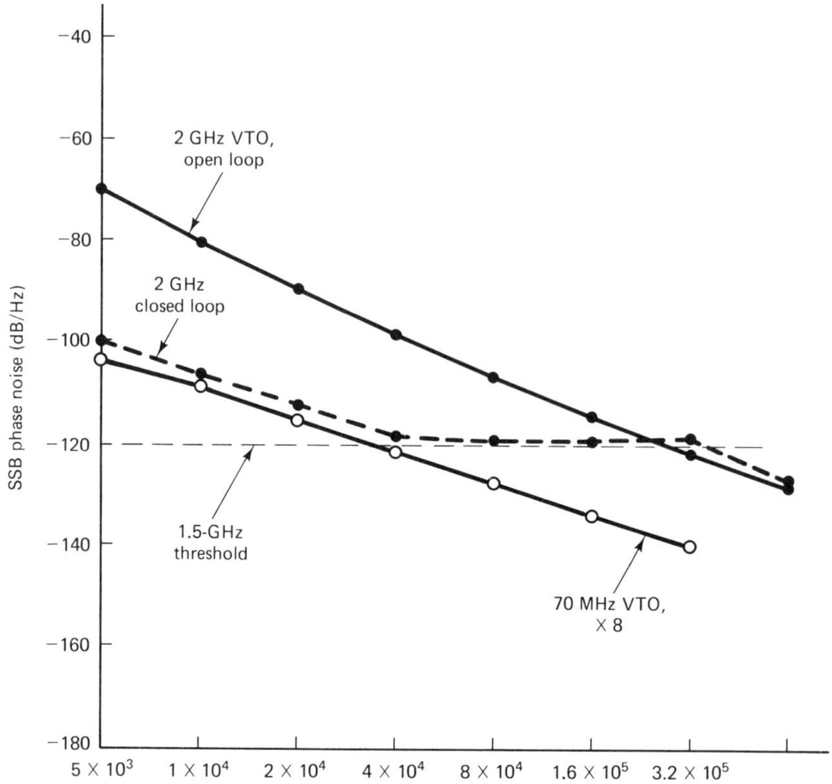

Figure 14-2. Synthesizer phase noise

to the 70-MHz reference SNR after times-8 multiplication as shown in Fig. 14-2. This condition is evident for f_m lower than 40 kHz, where the closed-loop, 2-GHz VTO phase noise is nearly equal to the 70-MHz, times-8 phase noise.

For values of f_m from 40 to 320 kHz and allowing for maximum measurement errors of 2 to 3 dB, the measured data are within a few decibels of the 1.5-GHz threshold. This threshold was the noise floor level at the output of the times-12 frequency multiplier. In the region between 40 and 160 kHz, the measured SNR is approximately 4 to 5 dB better than calculated; however, since each of the measured SNRs has a maximum error of ± 2 to 3 dB, this larger departure from predicted values can be expected.

REFERENCES

1. Payne, J. B., III, "Measure and Interpret Short-Term Stability," *Microwaves* (July 1976), pp. 34–45.
2. Fischer, M. C., "Analyze Noise Spectra with Tailored Test Gear," *Microwaves* (July 1979), pp. 66–75.

15

Direct and Indirect Combinations and Transformations

REFERENCE FREQUENCIES FOR BASE 2 SYNTHESIZER

This chapter begins with a description of techniques, mostly indirect, for obtaining the five reference frequency sources needed in the fast-switching, direct-type synthesizer diagrammed in Fig. 5-3 in Chapter 5. From a single 30-MHz crystal oscillator, the sources at frequencies of 600, 2948.8, 3000, 3450, and 3475.6 MHz are derived as illustrated in Figs. 15-1 and 15-2. Beginning with the 30-MHz standard of Fig. 15-1, the 600-MHz reference is obtained with a single loop containing a fixed divide-by-20 and a phase-frequency detector. A times-5 multiplier provides the simplest method to obtain the 3000-MHz reference.

The 3000-MHz reference is needed for the direct-type synthesis circuits and as a reference to the 3450-MHz source. This stable frequency is obtained from the VTO by downconverting to 450 MHz and then dividing by 15. This output is applied to a phase-frequency detector with a 30-MHz reference where locking to the 30-MHz standard is obtained. To derive the two remaining references at 2948.8 and 3475.6 MHz, similar

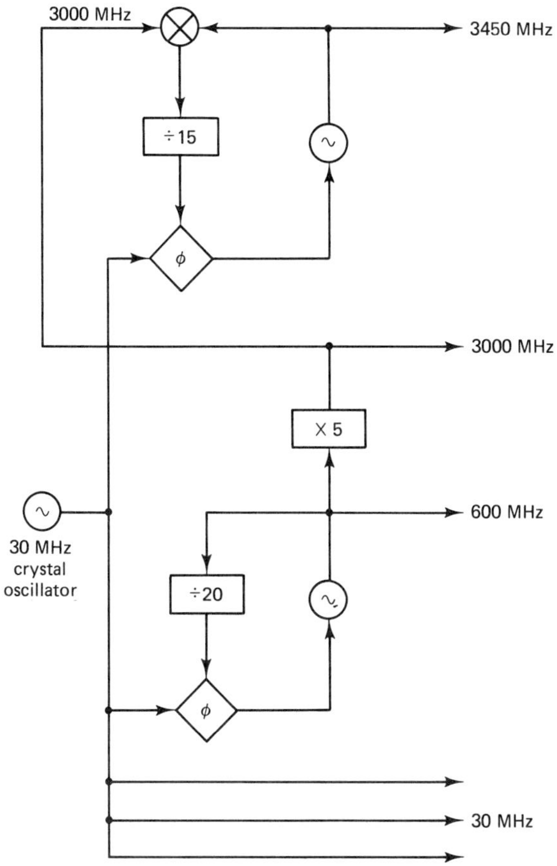

Figure 15-1. Synthesis of multiple phase-locked sources

converter and divider circuits are used as illustrated in Fig. 15-2. The two VTOs are phase-locked to the 30-MHz standard in nearly identical configurations, the only difference being divide-by-64 versus divide-by-128 and inputs of 3000 and 3450 MHz to the downconverters.

Briefly, the circuits needed to derive the five synthesizer references are uncomplicated in this design example; however, the distribution of these references within the synthesizer requires considerable planning to obtain a high degree of isolation from the mixers in the synthesizer. Whenever we have the condition where a fixed reference is applied to two or more mixers where the other inputs to the mixers are not at equal frequencies, all of the harmonic modulation products present in one mixer may be coupled to all other mixers through the paths provided by the fixed-

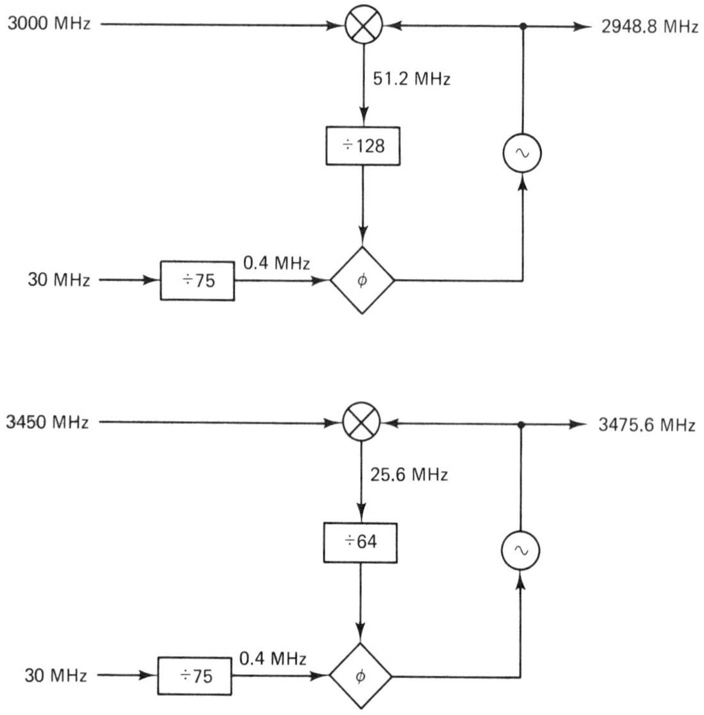

Figure 15-2. Synthesizing the 2948.8- and 3475.6-MHz sources

frequency inputs. To reduce this effect, the output from each source is initially RF-power-divided, and then each output line from the RF power divider must contain isolation devices to obtain the required degree of isolation from spurious signals generated within all of the synthesizer's mixers. These isolation techniques usually include a series of attenuators, amplifiers, or isolators.

REFERENCE FREQUENCIES FOR BASE 8 SYNTHESIZER

The base 8 synthesizer technique illustrated in Fig. 4-6 in Chapter 4 needs nine sources derived from a single standard. One way to accomplish this is shown in Fig. 15-3, where a crystal oscillator operating at 128 MHz provides a frequency reference to eight similar single-loop-locked oscillators; however, in this figure, only four of the phase-locked oscillators are shown. The remaining four oscillators have the following outputs, and

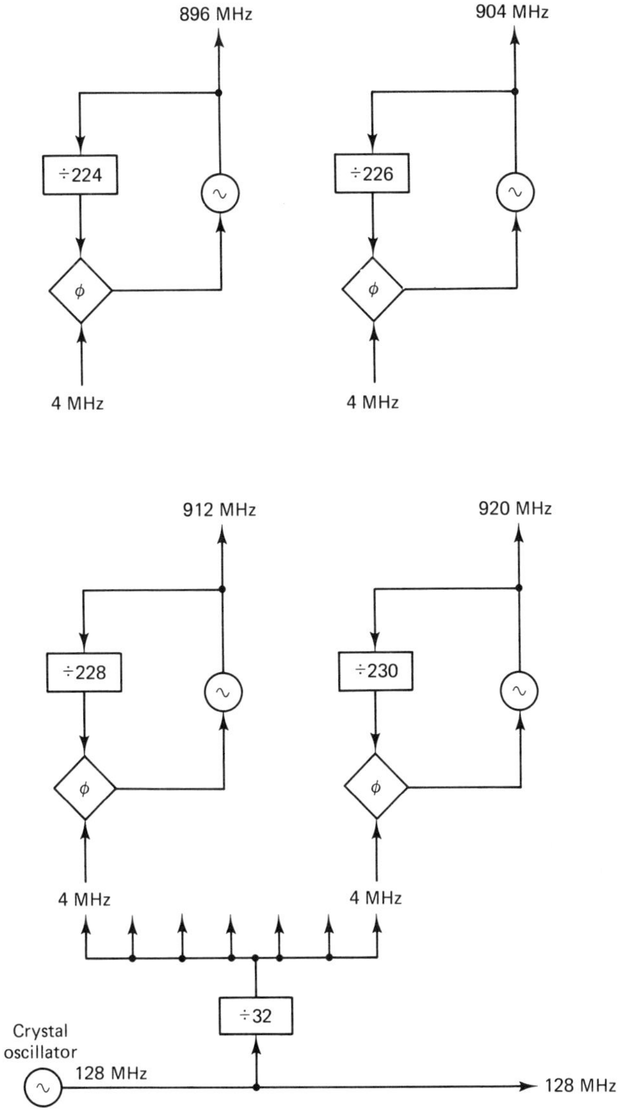

Figure 15-3. Synthesized references for base 8 synthesizer

the value of each feedback divider is listed for each loop:

928 MHz ÷ 232

936 MHz ÷ 234

944 MHz ÷ 236

952 MHz ÷ 238

REFERENCE FREQUENCY ISOLATION

Since each source in the band from 896 to 952 MHz of Fig. 4-6 is connected to three 1P8T RF switches, isolation devices are required to prevent spurious signals generated within each mixer from coupling to the other mixers.

Figure 15-4 is an example of some mixer spurious paths that exist if isolation techniques are not employed to reduce their detrimental effects. In this example, not all fixed-frequency sources are shown and not all inputs to the 1P8T switches are drawn; however, they would exist in an actual synthesizer. We assume the output frequency to be 1025 MHz; for this value, switches S0 and S2 are programmed to apply 896 MHz to mixers M0 and M2. The output of mixer M0 at 1024 MHz will be present at mixer M0's LO port at a power level of the IF output reduced by the

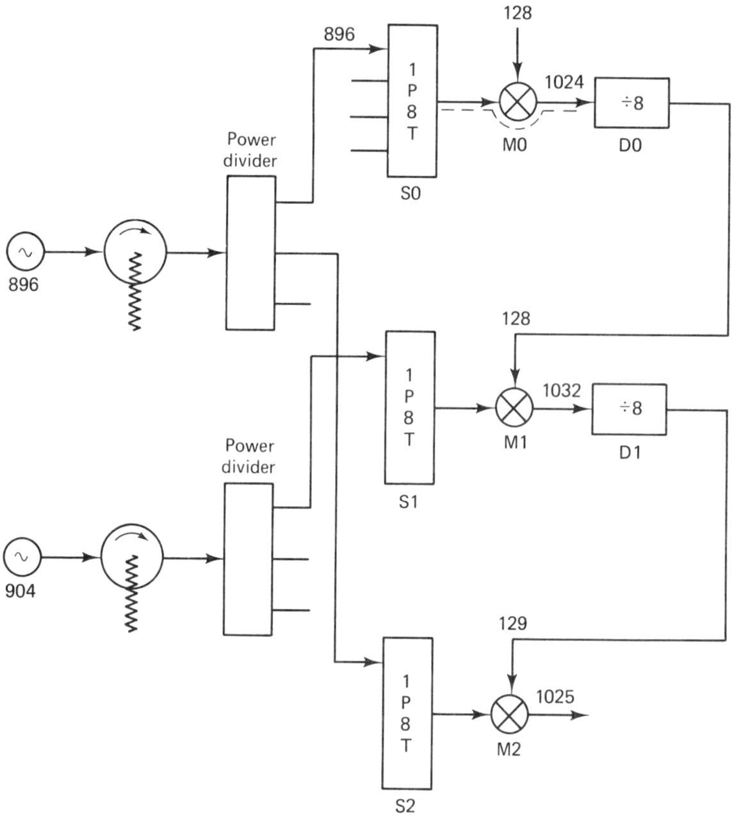

Figure 15-4. Mixer spurious signal paths

mixer's IF-to-LO-port isolation. Most microwave mixers will have on the order of 20- to 30-dB isolation for this condition. As the 1024-MHz spurious travels along the 896-MHz path, the next device in the path is the three-way power divider.

These devices, even when properly driven with an accurate 50-ohm source impedance, will usually provide only 20 to 30 dB of isolation at the power divider outputs; therefore, the 1024 spurious is present on the 896-MHz input to mixer M2's LO port on the order of 40 to 60 dB down from the mixer's output IF power level. The isolation of mixer M2's LO port to IF port is also in the range of 20 to 30 dB; thus the 1024 spurious will appear as a single sideband spurious to the 1025-MHz output at a power level of 60 to 90 dB down from the 1025-MHz level with the total isolation equal to the sum of the three devices' isolation: M0, M2, and three-way power divider.

There is the possibility that the spurious level would not even be 60 dB down if the LO of mixer M2 acts as an amplitude limiter and converts the single-sideband spurious signal to angular modulation where the signal to SSB ratio is 6 dB larger than the ratio of signal to single-amplitude spurious ratio. For our example, with a signal to single-amplitude spurious ratio at the LO port of mixer M2 of 40 dB, the resulting spurious phase sidebands would be 46 dB down (at 128 MHz removed from the 896-MHz carrier) and then transferred to the 1025-MHz IF at a frequency removed of 128 MHz. Thus, in the worst case, the IF output of M2 at 1025 MHz would have a single-amplitude spurious at 1024 MHz and at least 60 dB down and also a pair of phase-modulation spurious signals at 1025 ± 128 MHz and at least 46 dB down.

Returning to mixer M0 and considering it to be a normal mixer, the difference term of 896 − 128 MHz is also present at the IF output of mixer M0. This signal can also travel to the LO port of mixer M2 as just discussed. In addition, all of the lower-order intermodulation terms of M0 will also be present; for these reasons, any type of isolation technique to reduce the spurious present on the output synthesizer mixer must be effective over a wide bandwidth centered on the synthesizer mixer's LO port frequencies.

Various isolation techniques are diagrammed in Fig. 15-5. Beginning at the upper left, isolators can be added at the output of each of the eight three-way power dividers to gain an additional isolation of at least 20 dB (the usual minimum for these types of devices) from the levels discussed for Fig. 15-4. The devices also provide a termination to the LO source and can absorb the power reflections from switches that are not using that particular LO source as an LO to a mixer. The major disadvantage is that this technique requires three isolators per source for a total of 24 in a three-mixer, base 8 synthesizer. If the 1P8T switch dumps

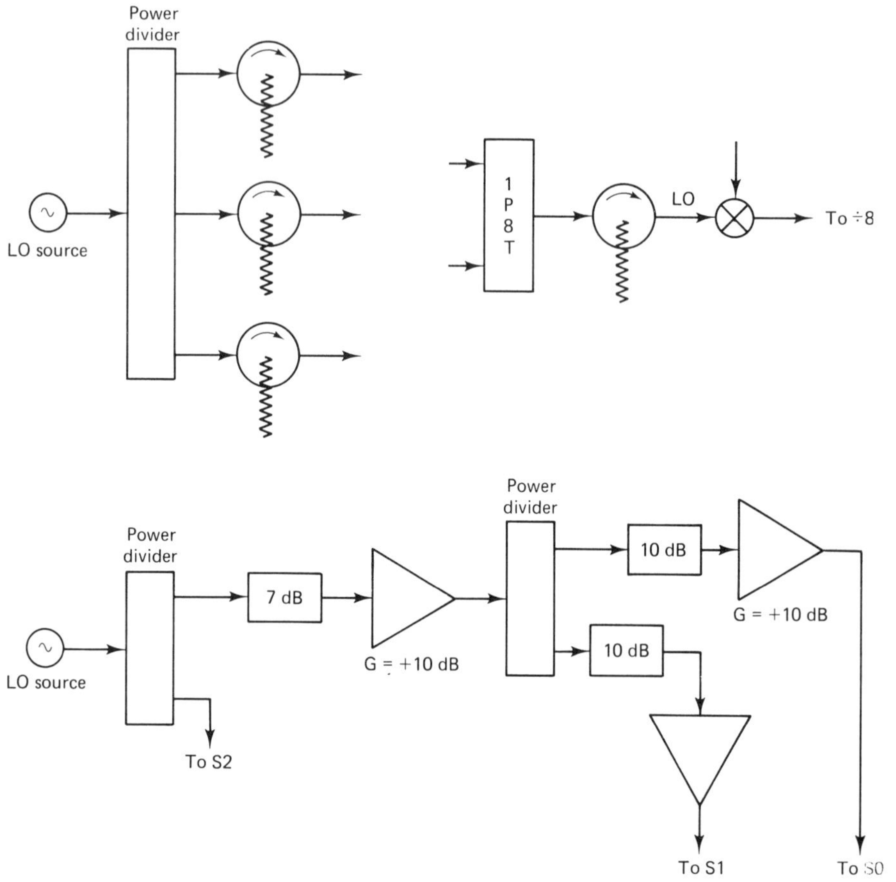

Figure 15-5. Mixer isolation techniques

unused input LO source power into a 50-ohm termination, a simpler isolation technique as shown at the upper right in Fig. 15-5 can be used.

With just one isolator for each of the three mixers, the isolation can be increased 20 dB over the levels of Fig. 15-4. It is easier to obtain even greater isolation by adding a second isolator to each mixer; thus more than 40 dB of isolation can be obtained with two isolators per mixer LO port. However, in order to use the simpler isolation technique, the 1P8T switch must terminate the seven unused inputs to prevent reflected power from pulling the phase-locked oscillators out of lock.

Since the low cost of active devices and low cost to manufacture microstrip circuits is a mature technology, active isolation techniques may cost considerably less than isolators and provide a higher level of isolation

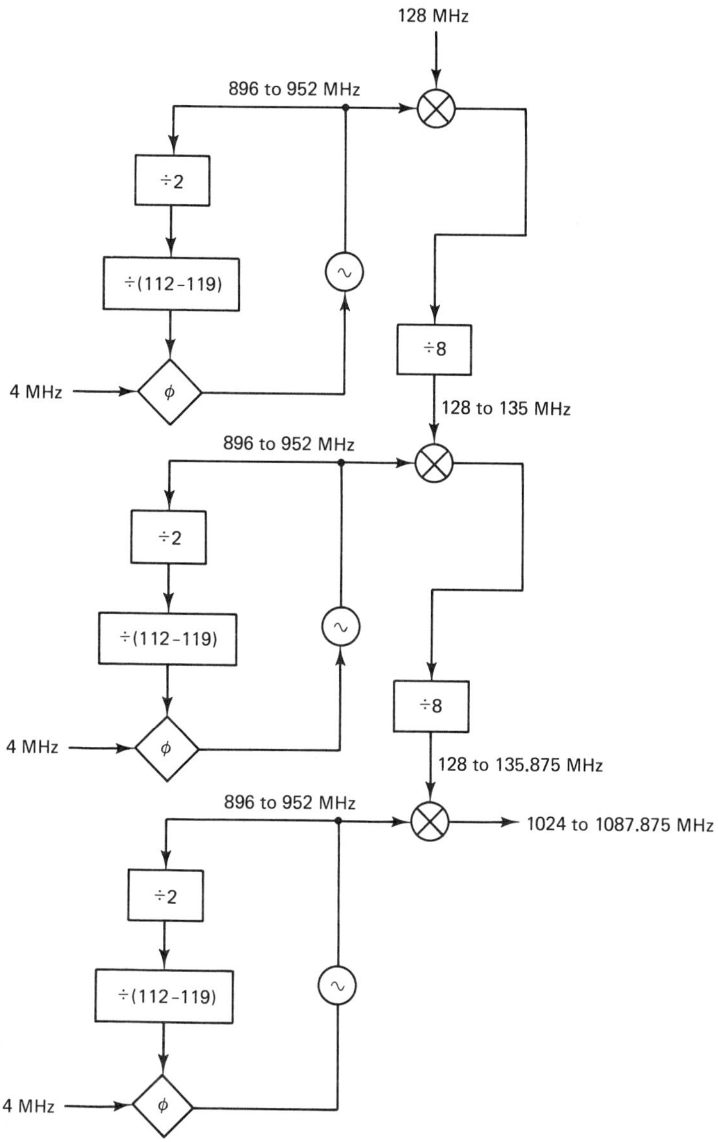

Figure 15-6. Base 8 synthesizer with agile phase-locked LO's

as shown in the lower half of Fig. 15-5. This technique takes advant of the requirement that highest isolation is needed at the output mi where no frequency dividers are present. As more frequency divisio present between a mixer's output and the synthesizer's output, isolat can be reduced because frequency divisions lower sideband spurious lev

Assuming an LO source output power level of +13 dBm, the t way power divider outputs are at +10 dBm, and one of these output connected to switch S2. The other output is attenuated by 7 dB and tl amplified by 10 dB to supply +13 dBm to the second two-way po divider; therefore, switches S1 and S0 will each receive +10 dB of source power. The isolation from S0 to S2 is provided by the reve isolation of two amplifiers, 17 dB of attenuation, and the two-way po dividers. With each amplifier having an isolation of 20 dB, this technic provides more than 57 dB of additional isolation than shown in Fig. 15 The path from S1 to S2 provides the same isolation as from S0 to S2.

Isolation from switch S0 to switch S1 does not need to be as large from S0 to S2 since the mixer connected to switch S1 contains a divi by-8 stage and will reduce phase-modulation spurious by 18 dB and a plitude spurious by 18 + 6 = 24 dB. The isolation from S0 to S1 provided by the reverse isolation of the one stage (10 dB of gain) amplif and the 10-dB attenuator and the two-way power divider. If more isolati is required, an additional 10-dB attenuator and single-stage amplifier c be placed in the RF path to switch S0.

TRANSFORMATION FROM DIRECT- TO INDIRECT-TYPE SYNTHESIZER

For the base 8 synthesizer of Fig. 4-6, the primary reason to select tl approach is to achieve very fast frequency agility with frequency switchi speeds of less than a microsecond. There are other applications whe frequency switching speeds of a few hundred microseconds are acceptab For these less demanding designs, the synthesizer in Fig. 4-6 can be si plified to the configuration diagrammed in Fig. 15-6. In this techniqu each 1P8T switch is replaced with a single VTO and a programmat divider, division from 112 to 119, to make an agile synthesized LO f each mixer. The configuration reduces the number of phase-locked VT from eight to three and deletes the requirement for three 1P8T RF switche in addition, isolation devices are no longer needed since each mixer L is derived from a single agile source.

The reduction of parts is achieved by combining direct and indirect techniques to optimize the technical design to meet specific tuning speed requirements. For this design with the 4-MHz reference for the single loops, we can approximate the loop bandwidth as 1/20 of this reference and as equal to 200 kHz. With the loop optimized for fast tuning, we can expect phase lock to occur in a time approximately equal to 5 μs (1 ÷ 200 kHz), and an additional 10 to 30 μs are needed for the phase transient, after lock occurs, to settle to the steady-state value.

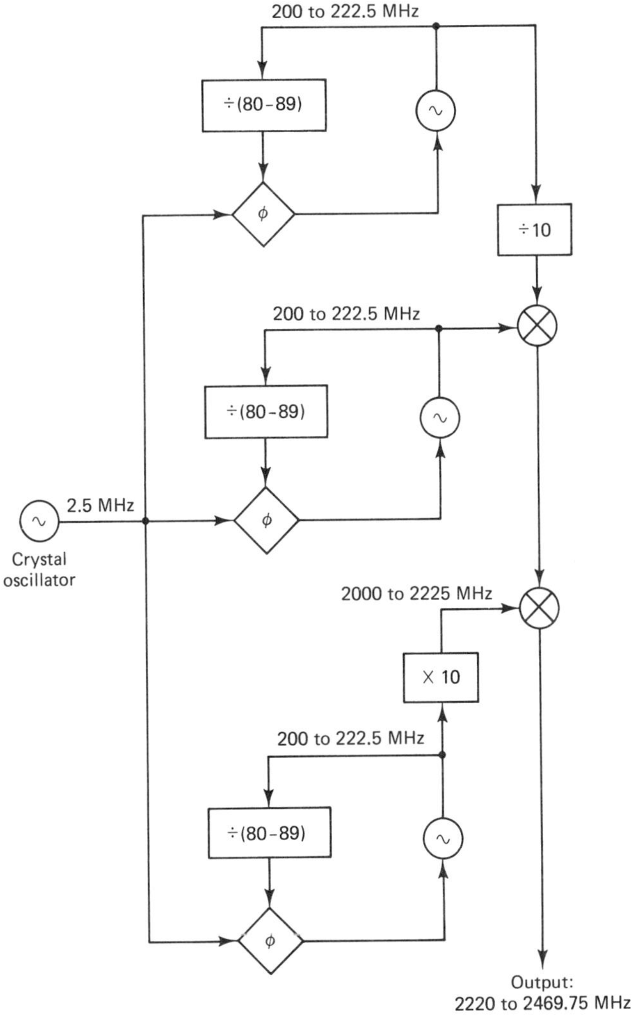

Figure 15-7. Base 10 synthesizer with agile phase-locked LOs

Since direct-type base 8 and 10 synthesizer techniques are very similar, we can modify the base 10 design in Fig. 5-1 of Chapter 5 in the same manner as we did the base 8 design of Fig. 5-3. The modified design is diagrammed in Fig. 15-7, where three agile phase-locked VTOs have replaced the ten oscillators, ten three-way power dividers, and three 1P10T RF switches. This configuration is no longer capable of achieving frequency switching times under 1 μs; however, the phase-locked-loop band-

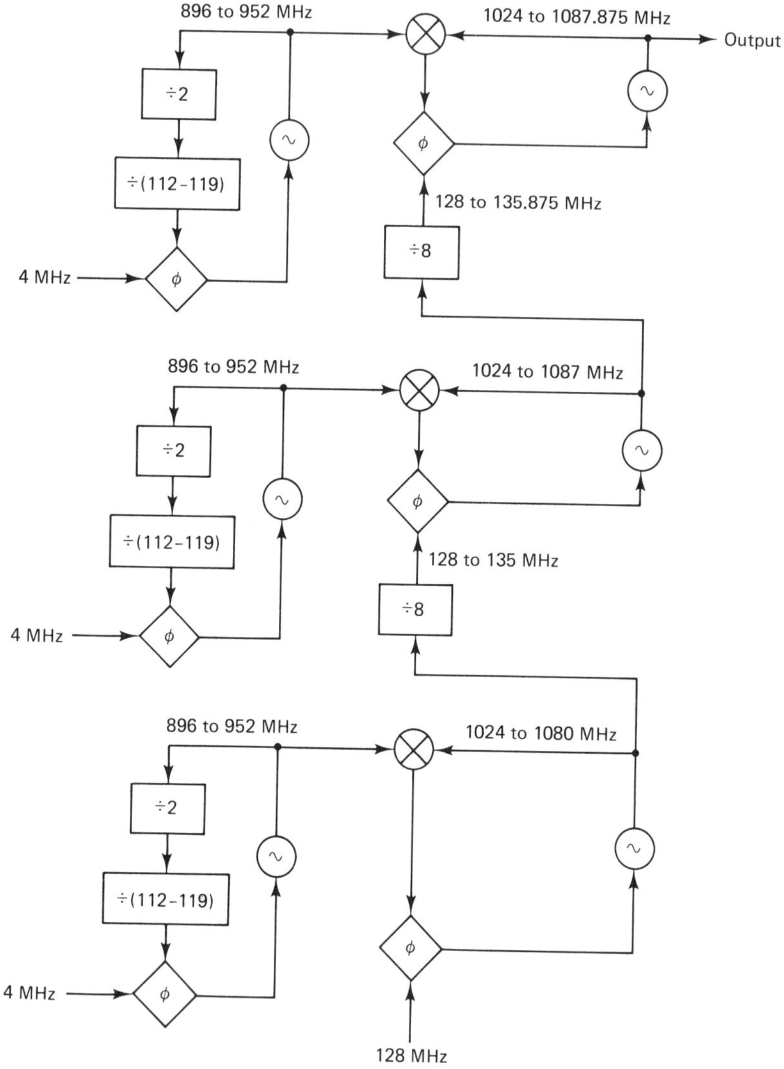

Figure 15-8. Transformation from direct-to indirect-type synthesizer

widths can have an upper limit of approximately 150 kHz. This bandwidth will allow the loop to achieve phase lock in under 20 μs, including the time required to position the VCO near its lock frequency. Allowing several time constants for the phase transient to settle out, a total frequency set-on time of 50 μs can be obtained.

The design in Fig. 15-6 is a combination of direct and indirect synthesizer techniques derived from the direct-type technique of Fig. 4-6. We can make the transformation to an indirect design by replacing the mixers with tracking loops as shown in Fig. 15-8; the three tracking loops are on the right side of the block diagram, and the three agile LOs are on the left side. Since the tracking loop phase-detector reference frequency is never lower than 128 MHz, the loop bandwidth can be very large, on the order of 6 to 10 MHz. Since this loop's speed will be much faster than the agile LOs' tuning speed, the agile LOs are the major contributor to limiting frequency set-on time at the synthesizer output of 1024 to 1087.875 MHz.

It is interesting to note the similarity of Fig. 15-8 to Fig. 9-2 in Chapter 9, where the indirect-type technique for base 10 was described. Here we initially examined designs of the direct type, transitioned these designs to a mixture of direct and indirect, and finally made the transition to an indirect type. In Chapter 16 we will present wideband microwave designs that are a mixture of direct and indirect techniques with very fast synthesizer agility as the major design requirement.

16

Combined Techniques for Maximum Performance

The main topic of this chapter is a synthesizer design technique where tuning speeds faster than 50 μs can be obtained in a wideband design extending from 500 MHz to 18 GHz. To reduce the total number of parts in an effort to lower cost, direct and indirect techniques are combined to achieve a design optimized for fast-tuning performance with reduced hardware complexity.

TRACKING LOOP

Figure 16-1 is a block diagram of the overall technical design. The synthesizer output frequency f_o extending from 2 to 4.5 GHz is obtained from a tracking loop where outputs from linearized VTO1, tunable from 2 to 3 GHz, and VTO2, tunable from 3 to 4.5 GHz, are selected for their part of the frequency band by the 1P2T switch. Note that this microwave switch is controlled by input control lines and is assumed to switch in less

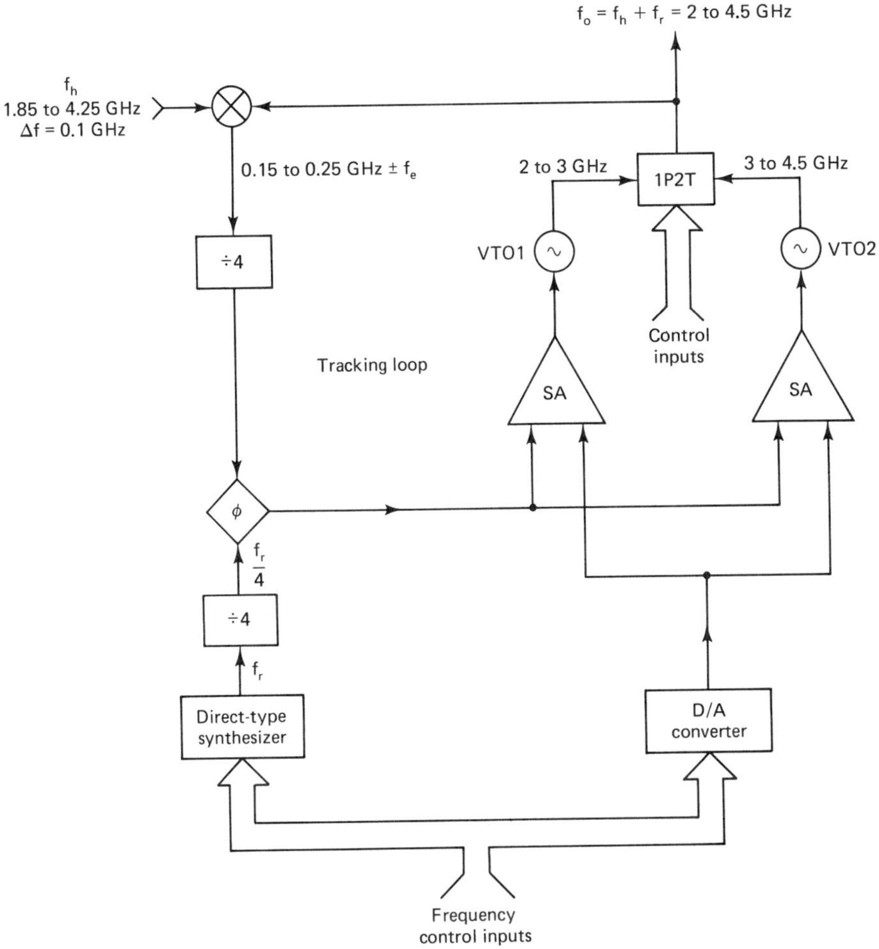

Figure 16-1. Technical approach for wideband microwave synthesizer

than 1 μs. Inputs to the tracking loop are reference frequencies f_h and f_r, where f_h is tunable in 100-MHz increments from 1850 to 4250 MHz, to be discussed later in this chapter.

The tracking loop oscillators, VTO1 and VTO2, are initially tuned by the D/A converter to a new programmed frequency value with an assumed frequency set-on accuracy to make the error to the phase detector, $\pm f_e/4$, less than the loop bandwidth. This is a critical requirement for the tracking loop since it must rapidly relock each time the synthesizer is programmed to a new frequency value. Methods to minimize initial set-on error include the use of linearized VTOs combined with a digital memory containing a look-up table to account for VTO linearization errors or placing a wideband discriminator loop around each VTO. A part of the VTO output is coupled to the mixer where the high-frequency reference f_h is also an input to the mixer.

The IF output from the double-balanced mixer extends from 150 to 250 MHz when the tracking loop is locked and contains an initial error frequency f_e each time the synthesizer is programmed to a new output frequency. Once locked, the IF frequency is equal to the direct-type synthesizer output frequency, f_r. The divide-by-4 circuits are needed to lower these two inputs to the phase-frequency detector since this device typically has a maximum operating limit of less than 100 MHz.

PHASE-DETECTOR REFERENCE FREQUENCY

The direct-type synthesizer providing reference frequency f_r is derived from the base 8 technique described in Chapter 4 with reference to Fig. 4-5. The octal design is selected to minimize the number of "add and divide" stages within the synthesizer. Figure 16-2 is a block diagram of this direct-type design; starting at the lower right, a 150-MHz input and a variable 1050- to 1094.8-MHz input (6.4-MHz increments) from the 1P8T switch S1 are upconverted by mixer M1 to 1200 to 1244.8 MHz. The divide-by-8 lowers this value to 150 to 155.6 MHz, and the iterative process continues at mixers M2, M3, and M4.

The output of mixer M4 is defined as

$$f_r' = 8 \times 150 + N_3 \times 6.4 + \frac{N_2 \times 6.4}{8} + \frac{N_1 \times 6.4}{8^2} + \frac{N_0 \times 6.4}{8^3} \qquad (16\text{-}1)$$

and was established in Fig. 4-5. In the current example, N_3 is the switch position of S4; N_2 is the switch position of S3; N_1 is the switch position of S2; and N_0 is the switch position of S1. The values of N_i are 0, 1, 2, 3, 4, 5, 6, and 7, where i has values of 0, 1, 2, and 3. Returning to Eq. (16-1), the product of $N_3 \times 6.4$ controls the output in 6.4-MHz increments; the product of $N_2 \times 6.4 \div 8$ controls the output in 800-kHz increments; the product of $N_1 \times 6.4 \div 64$ controls the output in 100-kHz increments; and the product of $N_0 \times 6.4 \div 512$ controls the output in 12.5-kHz increments. Therefore, the output from mixer M4 is tunable in 12.5-kHz increments from 1200 to 1251.1875 MHz. A total bandwidth of 100 MHz

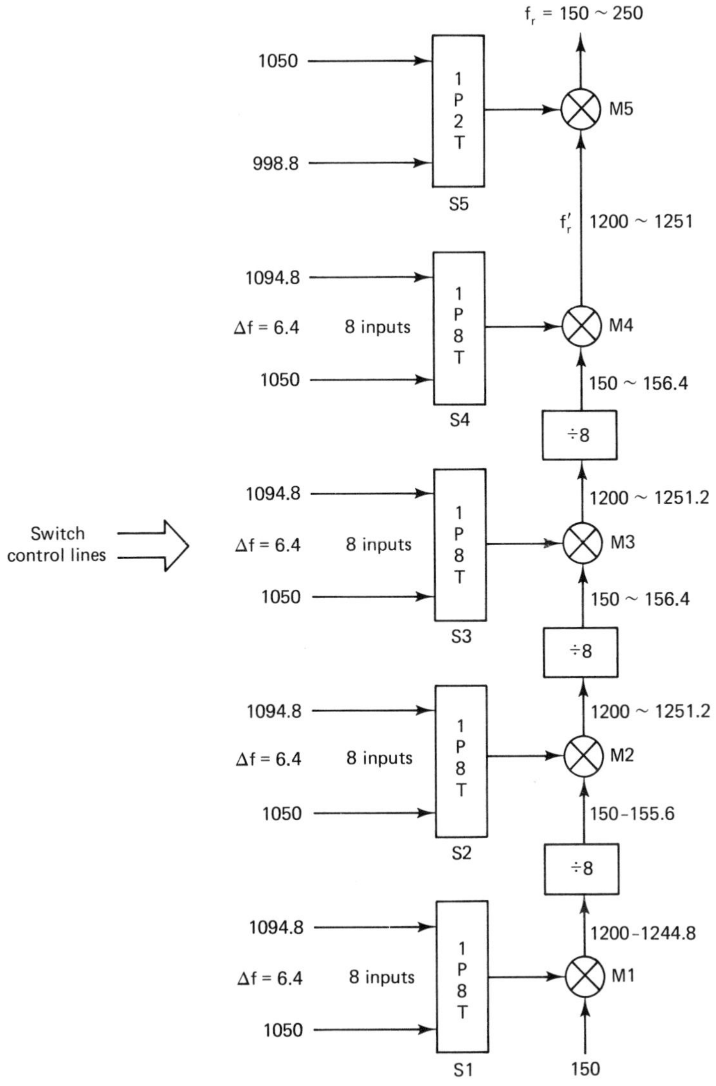

Figure 16-2. Phase detector reference frequency (f_r)

is required and is obtained with the addition of mixer M5, switch S5, and the source at 998.8 MHz.

With the LO of M5 at 1050 MHz, the output frequency f_r extends from 150 to 201.1875 MHz; lowering the LO to 998.8 MHz causes the output f_r to increase from 201.20 to 250 MHz. The time required to select any discrete value in these two bands is mainly limited by the 1P8T switch speed since bandwidths of the mixer IF filters are on the order of 60 MHz with filter delay times less than 50 ns.

MICROWAVE REFERENCE FREQUENCY

The microwave reference frequency f_h of Fig. 16-1 is obtained from a design similar to the one diagrammed in Fig. 12-3 in Chapter 12. This technique to synthesize reference f_h is illustrated in Fig. 16-3. As in the 2- to 4.5-GHz output tracking loop, two linearized voltage-tuned oscillators are required, each with an approximate bandwidth of 50%. Oscillator VTO3 operates from 1850 to 2850 MHz and is switched into the output at f_h in

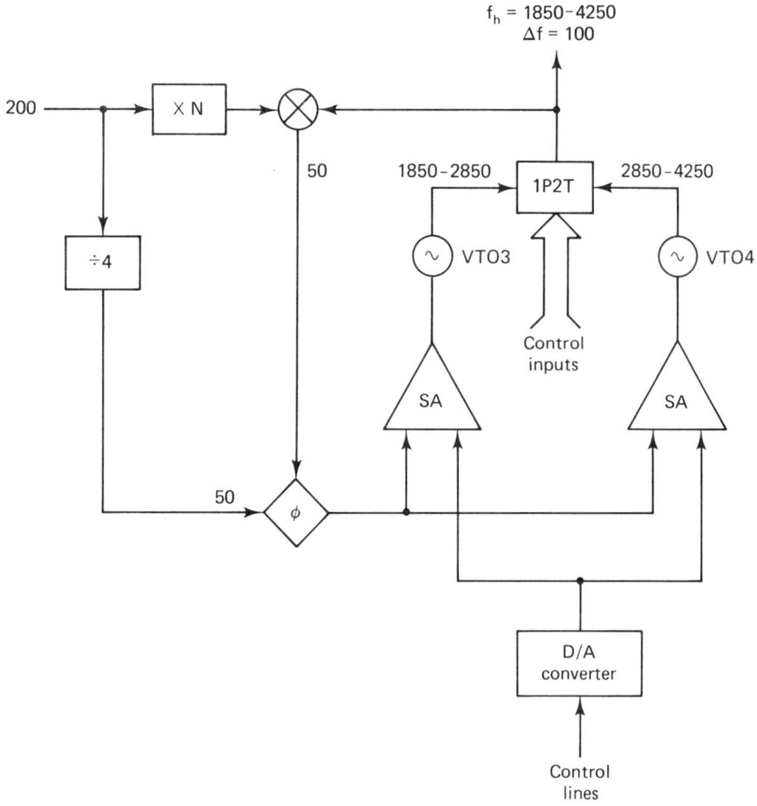

Figure 16-3. Microwave reference frequency (f_h)

parallel with VTO1 of Fig. 16-1. Oscillator VTO4 provides the signal at f_h when VTO2 of Fig. 16-1 is switched into the tracking loop.

For the configuration shown in Fig. 16-3, the available phase-locked frequencies start at 1850 MHz and extend to 4250 MHz in 100-MHz increments. Since the phase-frequency detector reference is 50 MHz, the maximum loop bandwidth will be on the order of 2 to 3 MHz; therefore, the time required to select any increment within the range of f_h should be well under 30 μs since the VTO can be tuned to new values in less than 20 μs, with phase locking under 1 μs and phase transients settled under several microseconds.

Returning to Fig. 16-1, the output of f_o is given by

$$f_o = f_h + f_r \qquad (16\text{-}2)$$

where reference f_h provides all tuning increments equal to or larger than 100 MHz; the smaller tuning increments are provided by f_r. Table 16-1 is a partial listing of f_o and is equal to the sum of f_h and f_r.

The basic techniques employed in this synthesizer design is to make reference f_r tunable under 1 μs by the use of direct frequency synthesis. Reference f_h is derived using indirect-direct techniques; however, it can be rapidly tuned in 100-MHz increments because the phase-frequency detector's reference is very high, 50 MHz, allowing for a wideband phase-lock loop. In addition, the tracking loop with output f_o also has a high reference to the phase-frequency detector; therefore, it can also be rapidly phase-locked to new input reference values.

TABLE 16-1

f_o (MHz)	f_h (MHz)	f_r (MHz)
2000	1850	150
2099.9875	1850	249.9875
2100	1950	150
2100 to 2199.9875	1950	150 to 249.9875
2200 to 2299.9875	2050	150 to 249.9875
2300 to 2399.9875	2150	150 to 249.9875
2900 to 2999.9875	2750	150 to 249.9875
3000 to 3099.9875	2850	150 to 249.9875
3100 to 3199.9875	2950	150 to 249.9875
3200 to 3299.9875	3050	150 to 249.9875
3700 to 3799.9875	3550	150 to 249.9875
3800 to 3899.9875	3650	150 to 249.9875
3900 to 3999.9875	3750	150 to 249.9875
4300 to 4399.9875	4150	150 to 249.9875
4400 to 4499.9875	4250	150 to 249.9875
4500	4250	250

BASE 8 SYNTHESIZER SOURCES

Frequency references required for the base 8 synthesizer of Fig. 16-2 can be derived from single-loop designs as illustrated in Fig. 16-4. In the top half of the figure, eight separate sources are required with a constant divide-by-N. The values tabulated for each N to achieve frequencies of 1050, 1056.4, 1062.8 MHz, and so forth, are listed at the upper right of this figure. Seven of the sources contain a four-way power divider to supply a signal to switches S1, S2, S3, and S4 of Fig. 16-2. The source at 1050 MHz requires a five-way power divider since an input at this frequency is needed at all four switches and switch S5. The other source needed at S5 is 998.8 MHz and is derived as illustrated in the lower half of Fig. 16-4.

The common input reference frequencies for all of the single-loop synthesizers are 0.4, 150, and 1000 MHz. An additional reference of 200

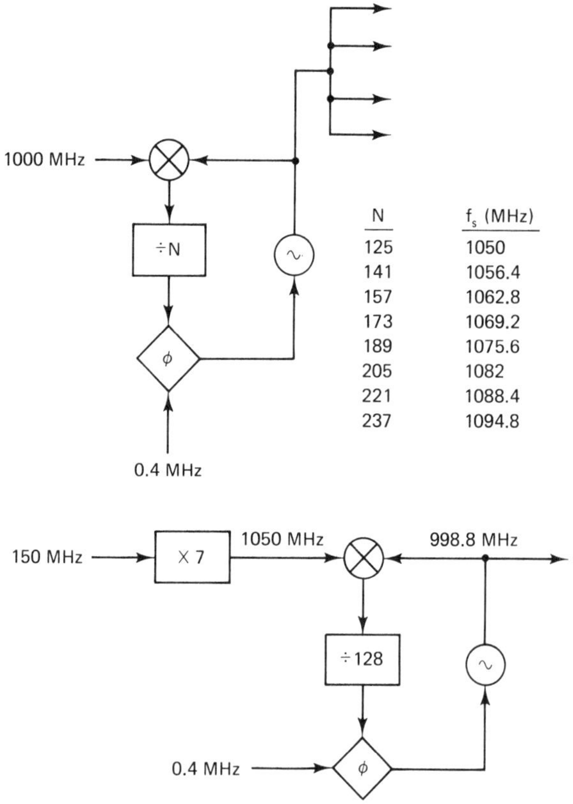

N	f_s (MHz)
125	1050
141	1056.4
157	1062.8
173	1069.2
189	1075.6
205	1082
221	1088.4
237	1094.8

Figure 16-4. Fixed frequency sources for the base 8 synthesizer

MHz is needed for the harmonic generator input of Fig. 16-3. The four frequencies can be derived directly from a single 50-MHz standard as follows:

$$0.4 \text{ MHz} = 50 \text{ MHz} \div 125$$

$$150.0 \text{ MHz} = 3 \times 50 \text{ MHz}$$

$$200.0 \text{ MHz} = 4 \times 50 \text{ MHz}$$

$$1000.0 \text{ MHz} = 20 \times 50 \text{ MHz}$$

Next, a single additional reference source is described to expand the synthesizer output from 500 MHz to 18 GHz.

CONVERTING TO A WIDER BANDWIDTH

Up to this point in the discussion, we have examined a 2- to 4.5-GHz synthesizer design containing direct and indirect techniques. Expanding the bandwidth to larger values will be described for the 0.5- to 2-GHz

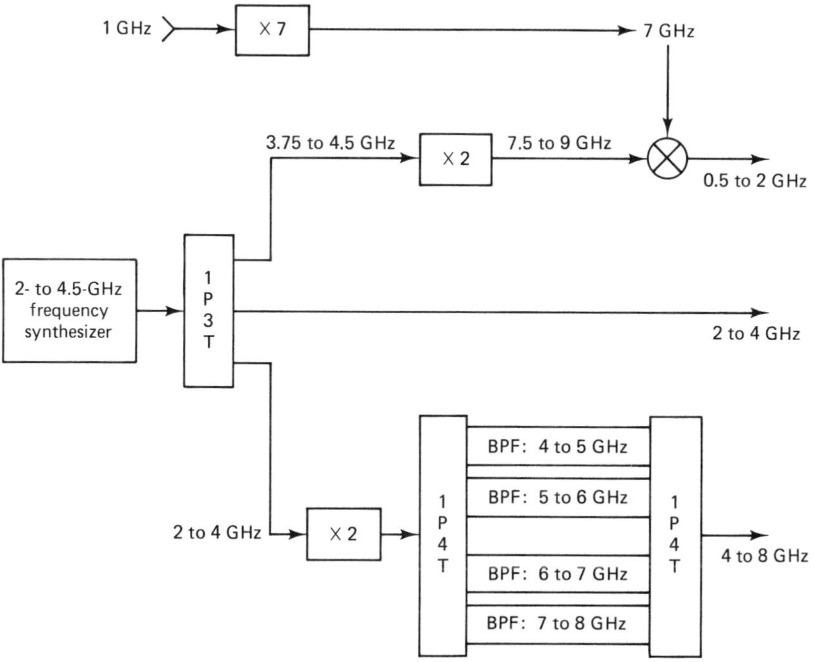

Figure 16-5. Expanding the synthesizer bandwidth

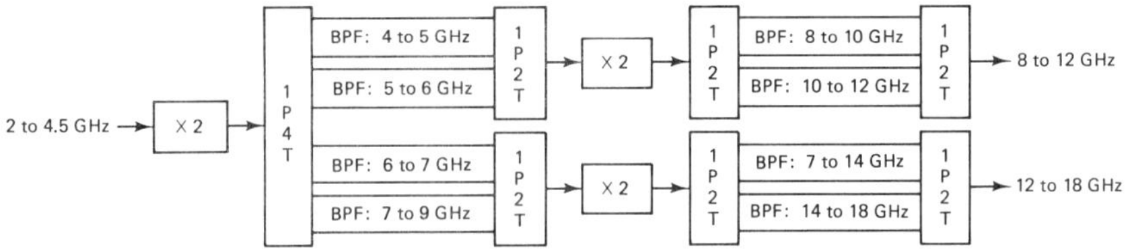

Figure 16-6. Increasing the output to 8 to 18 GHz

band, 4- to 8-GHz band, 8- to 12-GHz band, and 12- to 18-GHz band. The reader is reminded that our discussion is concerned only with the numerical part of the frequency synthesis; bandpass filter and amplifier requirements are not part of this discussion.

Figure 16-5 illustrates the major functions needed for the 0.5- to 2- and 4- to 8-GHz bands. From our 2- to 4.5-GHz synthesizer, a 1P3T switch directs the output to either frequency doubler or the 2- to 4-GHz output. The narrowband doubler at the top of the figure has an output extending from 7.5 to 9 GHz. This signal is downconverted by the DBM with a 7-GHz LO to the 0.5- to 2-GHz band.

In the lower part of Fig. 16-5, a wideband doubler's output is switched into four separate bandpass filters and recombined with an additional 1P4T switch to provide a single output containing 4 to 8 GHz. Since our basic 2- to 4.5-GHz synthesizer has a minimum tuning increment of 12.5 kHz, the frequency doubling associated with the 0.5- to 2- and 4- to 8-GHz bands will cause them to have a minimum tuning increment of 25 kHz.

For the higher bands at 8 to 12 and 12 to 18 GHz, it is necessary to add an additional doubling stage to the 4- to 9-GHz band as illustrated in Fig. 16-6. In this configuration, the 2- to 4.5-GHz output is doubled, switched, and filtered to a 4- to 6-GHz band and a 6- to 9-GHz band. Each of these synthesized frequencies is doubled, switched, and filtered to provide single outputs for the 8- to 12- and 12- to 18-GHz bands. Since the overall frequency multiplication is 4, the minimum tuning resolution in these two bands is 50 kHz. One final comment: The time required to change the output frequency from one value to another should be well under 1 μs since all devices in doublers and doubling paths are wideband with bandwidths exceeding 1 GHz.

The direct synthesis design of Fig. 16-2 will require isolation techniques to minimize spurious sidebands in the 150- to 250-MHz output. Mixers M4 and M5 are equally sensitive to LO input spurious since neither mixer's IF contains a frequency divider. In Fig. 15-5 in Chapter 15, the techniques described in the lower part of this figure can be implemented without a great deal of difficulty since all of the LO sources are in the lower microwave bands; thus the microstrip designs are easier to implement. Also, the amplifiers' reverse isolation can be very large if GaAs FETs are used as the single-stage amplifiers.

Frequency doubling was selected to obtain frequencies higher than the synthesized 1- to 4.5-GHz band because the resistive type of doubler is a wideband device with conversion loss on the order of 10 to 15 dB. After doubling and filtering, power levels can be restored with a two-stage gain block with approximately 20 dB of gain. Reactive doublers, using varactor diodes, are more power-efficient with lower conversion loss; however, these devices are difficult to stabilize in medium-bandwidth applications and work best in narrowband applications where only one stage of amplification is needed to restore power levels or to drive them at higher input power levels.

17

Synthesizers with Frequency Modulation

This chapter presents a description of techniques to add frequency modulation to indirect-type synthesizers. It begins with concepts developed in Chapter 7 on analog phase-locked loops as applied to a frequency-modulated reference source for a loop's phase detector. The techniques are then expanded to include synthesizers containing programmable dividers. In Chapter 18, techniques are presented to add angular modulation to direct-type frequency synthesizers. These techniques include the material developed in the current chapter as applied to direct synthesizers and the introduction of digital-type phase modulation, inherent in the base 2 and 4 direct-type synthesizer designs.

MODULATION WITHIN THE ANALOG TRACKING LOOP

From Chapter 7, Fig. 7-2 has been modified to the diagram of Fig. 17-1. For this loop, the input/output transfer function is given by

$$f_o = \frac{A_o}{1 + A_o} f_r \tag{17-1}$$

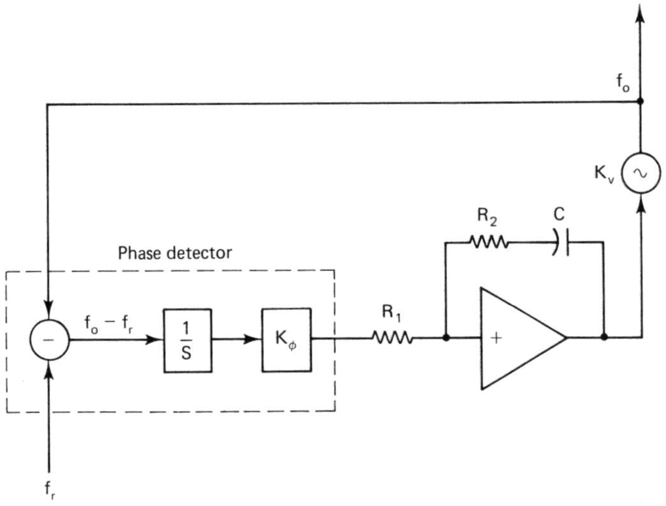

Figure 17-1. Modulation within analog loops

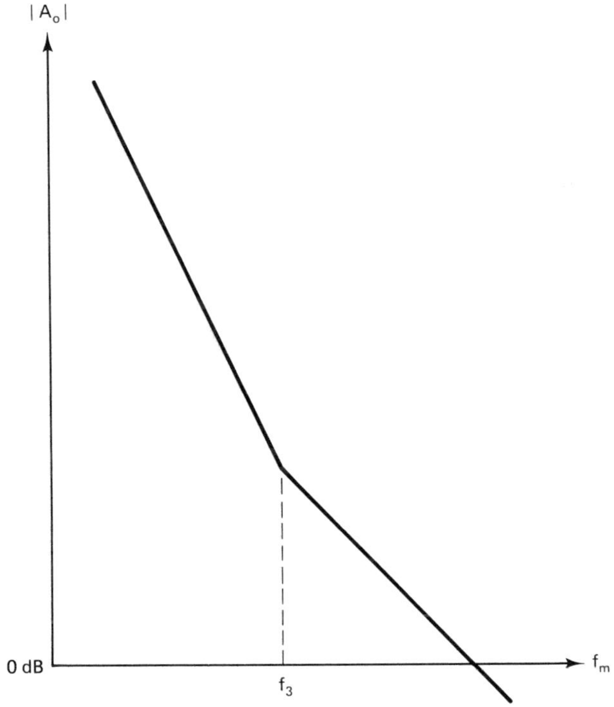

Figure 17-2. Open loop gain of the analog loop

where the open-loop gain A_o is

$$A_o = -j\left(\frac{R_2}{R_1}\right)\frac{K_v K_\phi}{f_m}\left(1 - j\frac{f_3}{f_m}\right) \tag{17-2}$$

and the low-pass filter bandwidth frequency f_3 is

$$f_3 = \frac{1}{2\pi R_2 C} \tag{17-3}$$

The frequency f_m is the rate of frequency modulation in the phase detector's reference f_r. The open-loop gain A_o is assumed to be a function of f_m as illustrated in Fig. 17-2.

For values of $f_m << f_3$, the open-loop gain is considerably greater than unity and is given by

$$A_o = (-1)\frac{R_2}{R_1}\left[\frac{K_v K_\phi}{f_m}\left(\frac{f_3}{f_m}\right)\right] \tag{17-4}$$

and the output frequency f_o is

$$f_o = \frac{A_o}{1 + A_o} f_r \approx f_r \tag{17-5}$$

Eq. (17-5) implies that the peak deviation of f_r at a deviation rate of f_m will be tracked by f_o when $A_o >> 1$. As rate f_m increases, the value of A_o decreases, and, depending on the loop's gain and phase margin, the peak deviation of f_o may exceed the peak deviation of reference f_r.

Returning to Eq. (17-1) with $f_3 >> f_m$,

$$f_o = \frac{-x}{1 - x} f_r \tag{17-6}$$

where x is equal to

$$\frac{R_2}{R_1}\left[\frac{K_v K_\phi}{f_m}\left(\frac{f_3}{f_m}\right)\right]$$

TABLE 17-1

x	20 log $-x/(1 - x)$ (dB)
10	0.91
20	0.45
40	0.22
100	0.09
200	0.04
400	0.02

The peak deviation of f_o will exceed the peak deviation of f_r according to Table 17-1. From the table we can see the importance of having a loop bandwidth considerably greater than the maximum deviation rate of reference f_r in order to have the peak deviation of f_o that is nearly the same as the peak deviation of f_r.

MODULATION WITH FREQUENCY DIVIDER IN THE LOOP

For single loops containing programmable dividers, concepts developed in Chapter 8 can be applied to frequency modulation in this type of loop. Figure 8-3 from Chapter 8 is shown in a simplified form in Fig. 17-3. For this configuration,

$$f_o = \frac{A_o}{1 + A_o/N} f_r \tag{17-7}$$

and the open-loop gain is

$$A_o = -j\left(\frac{R_2}{R_1}\right)\left(\frac{K_v K_\phi}{f_m}\right)\left(1 - j\frac{f_3}{f_m}\right) \tag{17-8}$$

where the value of f_3 is given by

$$f_3 = \frac{1}{2\pi R_2 C} \tag{17-9}$$

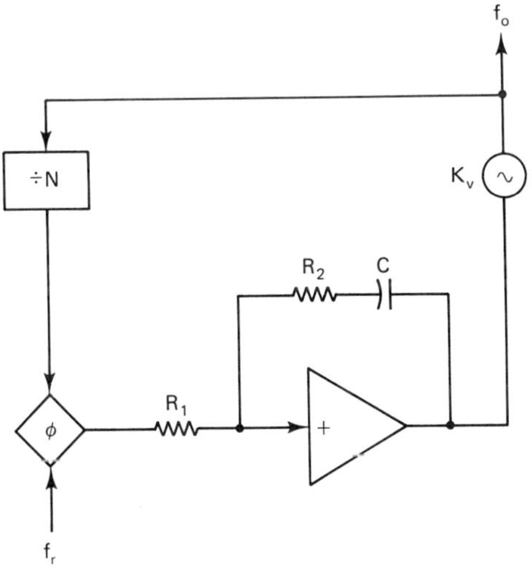

Figure 17-3. Modulation with frequency divider in the loop

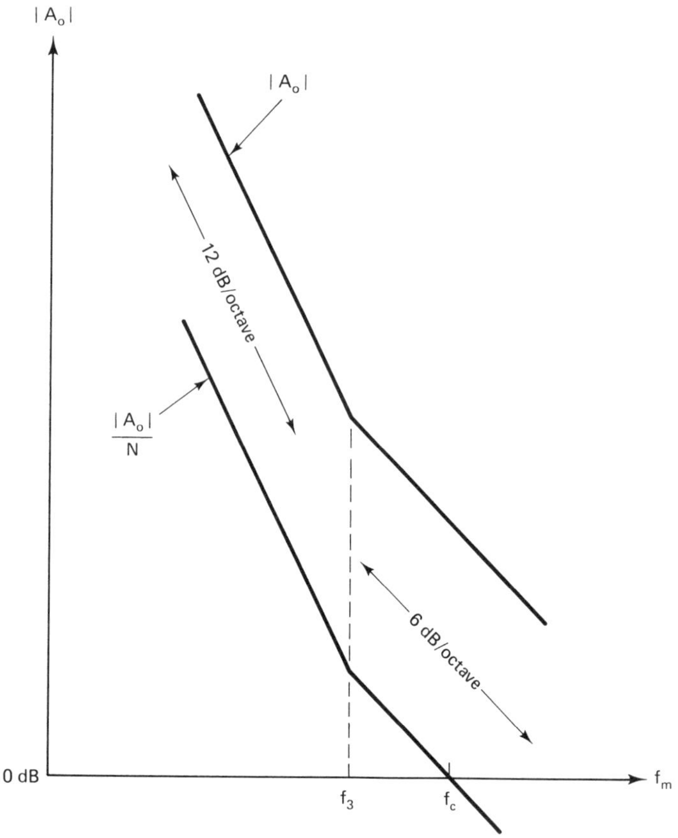

Figure 17-4. Open-loop gain of the loop with frequency divider N

The magnitudes of A_o and A_o/N are shown graphically in Fig. 17-4. For the loop to be stable, it is important for the value of $|A_o| \div N$ to be greater than 6 dB at $f_3 = f_m$. For values of f_m considerably less than f_3, the open-loop gain A_o is

$$A_o = -\left(\frac{R_2}{R_1}\right)\left(\frac{K_v K_\phi}{f_m}\right)\left(\frac{f_3}{f_m}\right) \tag{17-10}$$

Let x be as follows:

$$x = \left(\frac{R_2}{R_1}\right)\left(\frac{K_v K_\phi}{f_m}\right)\left(\frac{f_3}{f_m}\right) \tag{17-11}$$

Substituting this value into Eqs. (17-10) and (17-7), we get

$$f_o = \frac{-x}{1 - x/N} = \left(\frac{N}{1 - N/x}\right)f_r \tag{17-12}$$

TABLE 17-2

N/x	$20 \log \dfrac{1}{1 - N/x}$
0.001	0.009
0.002	0.017
0.004	0.035
0.01	0.087
0.02	0.12
0.04	0.35
0.1	0.91
0.2	1.94
0.3	3.1

Table 17-2 lists the excess peak deviation that is added to Nf_r.

Since x *is* equal to the absolute value of open-loop gain A_o, the value of divider N needs to be considerably smaller than A_o to have the peak deviation of f_o equal to $N \times f_r$. This is an important property of a loop containing a frequency divider for two reasons:

1. Incremental changes (peak deviation) of f_r are multiplied by N in the output frequency f_o; however, deviation rate (or the modulating frequency f_m) is not altered by divider N.
2. Since peak deviation is multiplied by N and deviation rate remains constant, the modulation index of f_o is N times greater than the modulation index of f_r.

We can also state this property as follows:

> The modulation index of frequency f_o is divided by N at the input to the phase detector.

We shall use this information in the next section.

MODULATION WITH RATES GREATER THAN LOOP BANDWIDTH

Maintaining the modulating frequency f_m (the rate of modulation) considerably less than the loop bandwidth, for any value of modulation index, assures the peak deviation of reference f_r to be imposed on f_o with $N =$

1. For values of N larger than unity, the peak deviation is multiplied by N.

To modulate the output f_o at rates much greater than loop bandwidth requires that the input modulation be a voltage applied directly to the output oscillator—preferably as V_{no} in Figs. 7-2 and 8-3 of Chapters 7 and 8, respectively. With the rate of modulation f_m considerably greater than loop bandwidth, the loop will be unable to cancel, entirely or partially, the modulation caused by V_{no}.

When no divider is present in the loop (this is identical to $N = 1$), the modulation index should not be allowed to exceed approximately 0.1 in order to prevent the oscillator from locking to a sideband of f_o. In Fig. 17-5(a) the modulation index is small and only one pair of sidebands is present. With the amplitude of $f_o >>$ either sideband amplitude, the loop will remain locked to f_o. As the modulation index is increased, more sidebands are present, and their amplitudes can exceed the amplitude of f_o. With a reduction in amplitude, the open-loop gain A_o is decreased

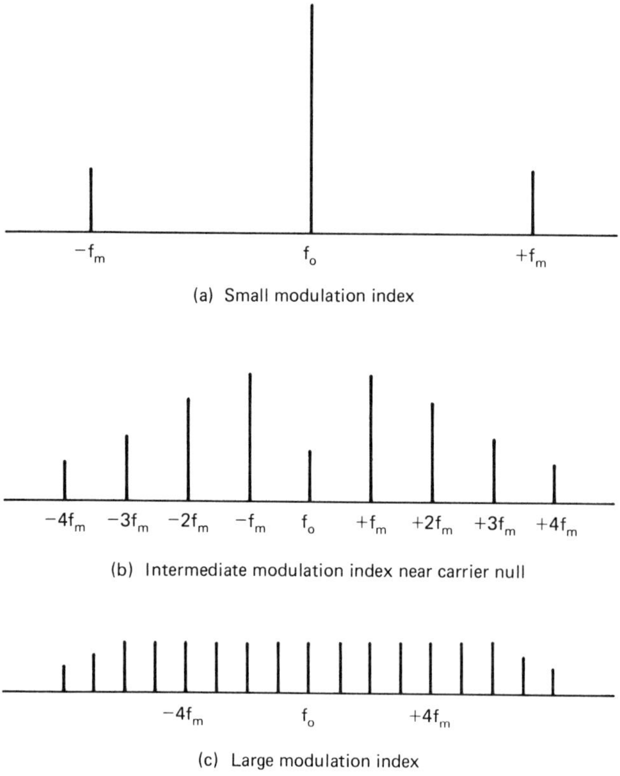

Figure 17-5. Frequency spectrum versus modulation index

and may cause A_o to have 0 phase margin when $A_o = 0$ dB! For these reasons, the modulation index at the phase detector must remain a small value.

When a divider is present in the loop, we can have a large modulation index at f_o with the modulation frequency greater than loop bandwidth for $N \geq 10M_o$ where M_o is the modulation index of the output f_o. With this limit on M_o, the modulation index of M_o/N is present at the phase detector and is less than 0.1; therefore, the loop will remain locked to reference f_r even at deviation rates greater than the loop's bandwidth when the output oscillator is modulated directly by an external voltage and not reference frequency f_r.

18

Adding Modulation to Direct-Type Synthesizers

In this final chapter, techniques for adding frequency modulation and/or phase modulation to direct-type synthesizers are presented. Recalling the direct synthesizer of Chapter 5 (Fig. 5-3) with an output extending from 600 to 702.3 MHz in 100-kHz tuning increments, the final output mixer stage is illustrated in Fig. 18-1(a). The mixer M10 and switch S10 are identical to components described in Chapter 5.

Switches S11 and S12 have been added to Fig. 18-1 in order to make output f_o selectable with or without frequency modulation, which may be required in frequency- or phase-modulated applications. To have phase-locked sources with FM capability, the techniques described in Chapter 15 (Fig. 15-2) are used as illustrated in Fig. 18-1(b). The converter at the left has inputs of 30 and 3000 MHz and outputs 2948.8 MHz. The converter at the right side has inputs of 30 and 2948.8 MHz and outputs 3000 MHz. In each of these converters, the 3000- and 2948.8-MHz inputs are fixed-frequency and do not contain any modulation. If the output modulation rates are considerably less than loop bandwidth, the 30-MHz reference is modulated, and no input modulation voltage is required. When the modulation rate is greater than loop bandwidth and the modulation

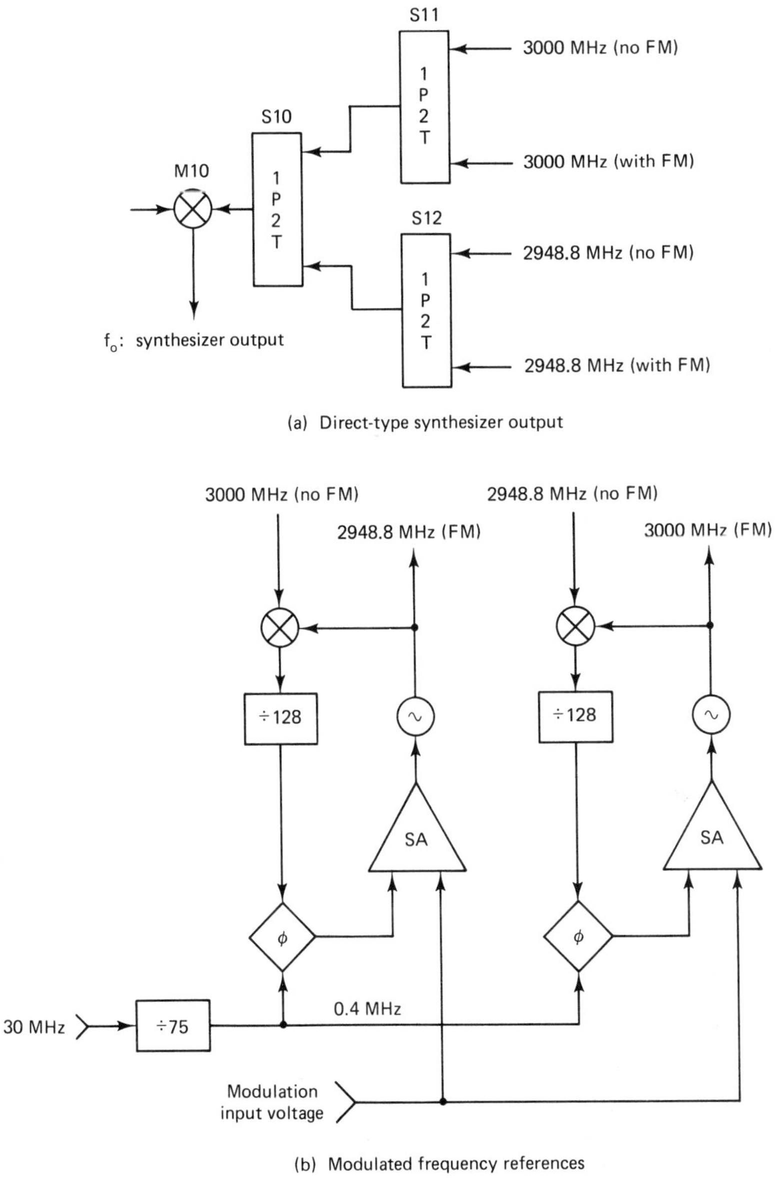

Figure 18-1. Adding frequency modulation to a direct-type synthesizer

index is less than 128/10, the VTOs are modulated directly with an external modulation voltage as shown in Fig. 18-1. The modulated phase-locked source outputs at 2948.8 and 3000 MHz are subject to all of the restrictions discussed in Chapter 17. Limiting the maximum modulation index to 0.1 at the phase detector requires that the modulation index be limited to 12.8 on the 2948.8- and 3000-MHz-locked VTOs.

If a larger modulation index is needed, the divide-by-128 frequency divider can be replaced with a larger value. Of course, the loop bandwidths will be reduced if larger dividers are added.

Discrete values of phase modulation (0, 90, 180, 270) or a linear voltage-controlled phase shifter can be inserted in the signal path between mixer M10 and switch S10 to add the phase-modulation capability. The advantage of placing the phase modulator in this path is the simpler circuit design where BW is approximately 1.7%. With this small bandwidth, it is possible to design more accurate phase modulators than with the larger 16% bandwidth signal path of the synthesizer output extending from 600 to 702.3 MHz. In addition, the time delay between application of modulation and the modulated output is minimized due to the proximity of this signal path to the synthesizer output.

Frequency shift-keying modulation is a natural part of the direct-type design since the output can switch to any frequency value in extremely short time intervals. The number of states in our example, shown in Fig. 5-3 in Chapter 5, is 1024, and this is equal to the number of selectable frequencies in the synthesizer output.

BINARY PHASE MODULATORS

Now phase modulators with a 0°/180° or 0°/90°/180°/270° capability are added to the direct-type synthesizer. The advantage of these designs is that the synthesizer output will contain accurate phase modulation with the smallest phase shift considerably less than 180°. Figure 18-2 is an example of one technique for a base 2 direct-type design where the divide-by-2 is used. This figure is similar to Fig. 4-9 in Chapter 4, except that it has been normalized to unity. We assume that the inputs, represented by the numeral 9 on the left side, have two frequency values (1000 and 1004 MHz in Fig. 4-9). In addition, we assume that the inputs on the right side, represented by the numeral 8, also have two frequency values (900 and 868 MHz in Fig. 4-9).

Beginning at the upper left, switch S1 is assumed to be electronically controlled to select either the 0 phase shift or the 180° phase-shifted signal.

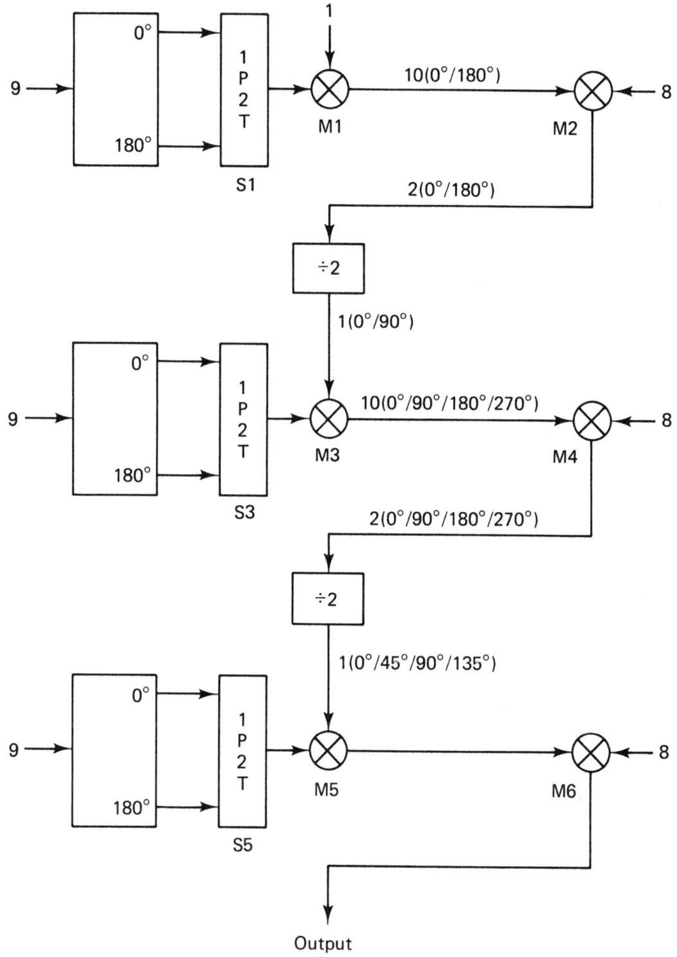

Figure 18-2. Adding phase modulation to a direct-type synthesizer

With the filter on the output of mixer M1 designed to pass the sum term, the output from M1 can be selected for either 0° or 180° phase shift. This phase information is downconverted at M2 and then divided by 2. Since a divide-by-2 is present, the phase shift will be divided by 2 and is equal to 0° or 90°. At mixer M3 output, the sum of 1 (0° or 90°) and 9 (0° or 180°) allows the four possible phase states (0°/90°/180°/270°) to be selected by the positions of switches S1 and S3.

The output of M4 downconverts this phase-shift information to the second divide-by-2, where phase shifts of 0°, 45°, 90°, and 135° are caused by the divider. In mixer M5, these four possible phase shifts can be combined with the 0°/180° from switch S5; therefore, the output phase shift can be selected to any value spaced 45° apart.

The major disadvantage of this approach is using a 1P2T switch to select the 0° or 180° phase shift from the 180° hybrid. Since the output signal amplitude from the switch will go to 0, the phase-shifted signals will contain amplitude modulation. To avoid this problem, modulators developed for QPSK communications systems using two double-balanced mixers and two 90° hybrids can be substituted into this design. These modulators have amplitude changes of 3 dB, and when combined with an amplitude limiter, the output amplitude would vary no more than a few tenths of a decibel.

Returning to Fig. 18-2, let us assume that we have deleted the hybrid-switch combination and have normalized frequency ("9") applied directly to mixers M1, M3, and M5. Now we place the constant amplitude 0°/180° phase shifters at the outputs of mixers M2, M4, and M6. For this configuration, we will have the same results as the original configuration: All phase changes on the left side of the Fig. 18-2 were up- and downconverted to the outputs of the mixers on the right side of the figure. We shall use this technique in our next example.

BASE 2 SYNTHESIZER INCORPORATING DIVIDE-BY-4

The synthesizer design illustrated in Fig. 5-3 is shown in a simpler diagram in Fig. 18-3 and includes the addition of the four-state phase shifters PS1, PS2, and PS3. The switches for selection of phase shift have been deleted in this figure; however, they would be present in an actual synthesizer. The fixed-frequency sources are shown in pairs to the LO port of each mixer where a single source is actually applied to the mixer by switches S1 through S10 in Fig. 5-3.

Returning to Fig. 18-3 and beginning with phase shifter PS1, the phase shift is assumed to be electronically selected for one of the four possible states: 0°, 90°, 180°, or 270°. In addition, the amplitude variations are assumed to be no greater than 3 dB at the phase-shifter output. When this signal, with four possible phase states, is applied to divider D4, two changes occur. The first is that the divider acts as an amplitude limiter; therefore, the output of divider D4 does not contain amplitude variations. The second change is the phase shifts are divide-by-4; thus the available phase states are 0°, 22.5°, 45°, and 67.5°.

The signal is next applied to mixer M7, where it is upconverted and then downconverted by mixer M8. After passing through phase shifter

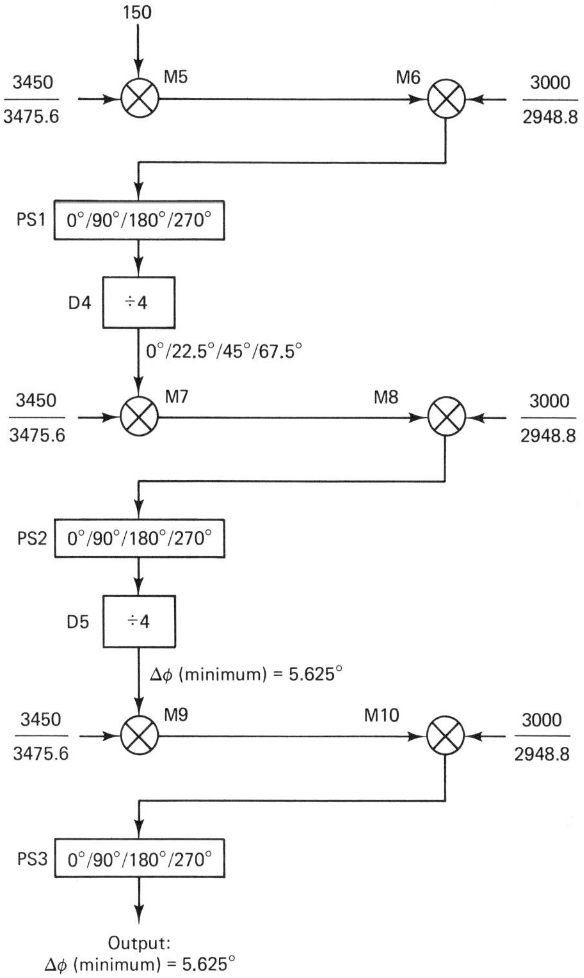

Figure 18-3. Base 2 direct synthesizer with phase modulation

PS2, where phase states of 0°, 90°, 180°, or 270° can be selected, 16 possible phase states are available every 5.625° from 0° to 84.375° after division by 4 of divider D5. In the final up- and downconverter pair of mixers, M9 and M10, the signal is applied to phase shifter PS3.

All of the input phase states to PS3 can be shifted by either 0°, 90°, 180°, or 270° at the final synthesizer output. This allows almost a full 360°—actually, 354.375°—of phase shift available within a minimum selectable shift of 5.625°. To enhance the capability of our synthesizer to include both frequency and phase modulation, we can add fixed sources with FM as inputs to mixer M10 as was illustrated in Fig. 18-1.

If the divide-by-4 circuits in the synthesizer IF are not resettable to 0 internal accumulated counts, phase coherency cannot be maintained with respect to the applied modulation rate. For example, if PS2 is changed from 0 phase shift to 90° phase shift when divider D5 has an accumulated count of either 1, 2, or 3, the divider will count down from this initial value before its output changes states; thus the accumulated counts act as a random delay prior to the final output at PS3 with the new phase information.

If the frequency dividers D4 and D5 can be reset to 0 accumulated counts each time phase shifter PS1 or PS2 is programmed to a new phase-shift value, the random delay can be eliminated if it is a problem. The delay through the synthesizer IF will then be a constant and equal to four cycles of the divider's output frequency; in addition, changes in phase states of PS3 are now coherent with the rate of applied phase modulation.

Phase shifters can be classified into two basic groups. In one group is the digital type, where diodes are used as switching elements to obtain discrete phase steps in multiples of 90°. References [1 through 4] describe the circuit design techniques used in the incremental type of phase shifters. The other group of phase shifters is the continuous, monotonic type, where the phase shift is usually derived from a varactor diode acting as a variable reactance on a short length of transmission line. References [5, 6, and 7] describe the theory and performance of the linear phase-shifter types.

REFERENCES

1. White, J. F., "Diode Phase Shifters for Array Antennas," *IEEE Transactions on Microwave Theory and Techniques*, MTT-22, no. 6 (June 1974), 658–74.
2. Andren, C., "PSK Sidebands Reduced by Premodulation Filtering," *Microwave Journal* (January 1978), pp. 69–73.
3. Moody, R., "Biphase and Quadriphase Digital Modulators," *Microwave Journal* (May 1982), pp. 160–62.
4. Shoko, R. K., and Arain, M. H., "L-Band Quadriphase Phase Shift Keyed Modulator," *Microwave Journal* (September 1982), pp. 131–38.
5. Garver, R. V., "360° Varactor Linear Phase Modulator," *IEEE Transactions on Microwave Theory and Techniques*, MTT-17, no. 3 (March 1969), 137–47.
6. Hopfer, S., "Analog Phase Shifter for 8–18 GHz," *Microwave Journal* (March 1979), pp. 48–84.
7. Bowman, G. A., "Linearize FM Transmitters over Wide Bandwidths," *Microwaves* (June 1979), pp. 66–75.

Appendix

CASE 1: DOWNCONVERSION

For a mixer with input frequencies f_l, and f_r, the downconverted output at frequency f_o is

$$f_o = f_l - f_r \tag{A-1}$$

Expressing this function as a ratio,

$$\frac{f_r}{f_l} = 1 - \frac{f_o}{f_l} \tag{A-2}$$

The interference frequency caused by harmonic mixing is

$$f_i = nf_l - mf_r \tag{A-3}$$

Note:

$$f_i = \pm f_o$$

Expressing Eq. (A-3) as a ratio,

$$\frac{f_i}{f_l} = n - m\frac{f_r}{f_l} \tag{A-4}$$

Substitute Eq. (A-2) into (A-4):

$$\frac{f_i}{f_l} = n - m\left(1 - \frac{f_o}{f_l}\right) \tag{A-5}$$

Rearrange Eq. (A-5):

$$\frac{f_i}{f_l} = n - m + m\left(\frac{f_o}{f_l}\right) \tag{A-6}$$

Rearrange Eq. (A-6):

$$m\left(\frac{f_o}{f_l}\right) - \frac{f_i}{f_l} = m - n \tag{A-7}$$

Factoring (f_o/f_l):

$$\frac{f_o}{f_l}\left[m - \left(\frac{f_l}{f_o}\right)\left(\frac{f_i}{f_l}\right)\right] = m - n \tag{A-8}$$

For f_i, substitute $f_i = \pm f_o$:

$$\frac{f_o}{f_l}(m \pm 1) = m - n \tag{A-9}$$

Rearrange Eq. (A-9):

$$\frac{f_o}{f_l} = \frac{m - n}{m \pm 1} \tag{A-10}$$

Substitute Eq. (A-10) into (A-2):

$$\frac{f_r}{f_l} = 1 - \frac{m - n}{m \pm 1} \tag{A-11}$$

Limiting the ratio of f_r/f_l to values greater than 0.5 and less than 1 causes Eq. (A-10) to be less than 0.5. We can compute Eq. (A-11) for all combinations of the integers n and m that cause an interference at f_o/f_l. The interference values are points on the line in Fig. A-1. The straight line is given by Eq. (A-2) with the limitations imposed as stated. Each point is accompanied by notations; for example,

3/5: (2,4) (4,6) (5,9)

and is

f_r/f_l: (n,m) (n,m) (n,m)

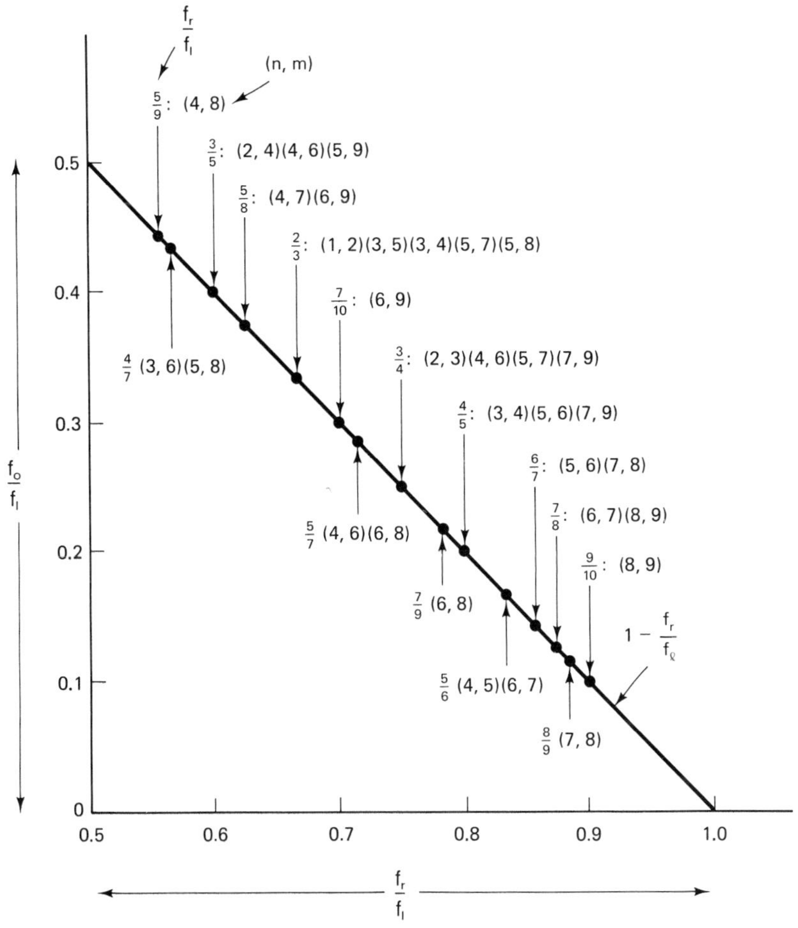

Figure A-1 Downconverter with output less than either input

where 3/5 is the ratio of f_r/f_l that will cause harmonically generated spurious interference from the combinations of

n	m
2	4
4	6
5	9

The points on this line in Fig. A-1 are for all combinations of n and m from 1 to 9. Inspection of this graph reveals that most points on the line for ratios of f_r/f_l between 5/9 and 4/5 have low-order harmonics of either n or m. For values of f_r/f_l greater than 6/7, the order of the harmonics is much higher; therefore, the interference power level should be very small compared to the power level of f_o.

CASE 2: DOWNCONVERSION

Using the equations developed for Case 1, we next limit the ratio of f_r/f_l to greater than 0 and less than 0.5, as illustrated in Fig. A-2. From

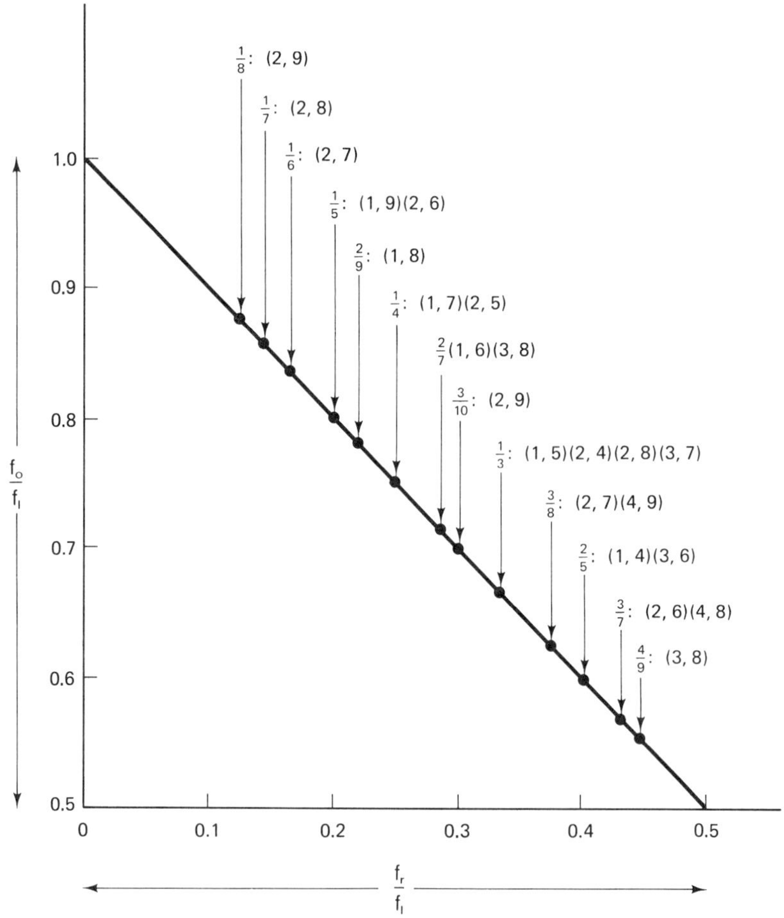

Figure A-2 Downconverter with output less than one input

Eq. (A-11),

$$\frac{f_r}{f_l} = 1 - \frac{m - n}{m \pm 1}$$

Since f_r/f_l is less than 0.5, the value of $(m - n)/(m \pm 1)$ is restricted to greater than 0.5 and less than 1. We can now compute the fractional values of f_r/f_l that cause an interference in the ratio of f_o/f_l as indicated by the points in the line in Fig. A-2. *Note:* These points are for any combination of n and m for values of n and m from 1 to 9.

We note on the line the absence of interference points when f_r/f_l is less than 0.1, and with this restriction on f_r/f_l, we have an excellent lower sideband upconverter.

CASE 3: UPCONVERSION

For an upconverter with output frequency f_o and inputs at f_l and f_r, the output is given by

$$f_o = f_l + f_r \tag{A-12}$$

Rearrange to

$$\frac{f_r}{f_l} = \frac{f_o}{f_l} - 1 \tag{A-13}$$

The interference frequency equal to f_o is

$$f_i = nf_l - mf_r \tag{A-14}$$

Rearrange Eq. (A-14):

$$\frac{f_i}{f_l} = n - m\frac{f_r}{f_l} \tag{A-15}$$

Rearrange Eq. (A-15):

$$\frac{f_i}{f_l} = n - m\left(\frac{f_o}{f_l} - 1\right) \tag{A-16}$$

Rearrange Eq. (A-16):

$$\frac{f_i}{f_l} + m\frac{f_o}{f_l} = n + m \tag{A-17}$$

Factor f_o/f_l:

$$\frac{f_o}{f_l}\left[\left(\frac{f_i}{f_l}\right)\left(\frac{f_l}{f_o}\right) + m\right] = n + m \tag{A-18}$$

Substitute $f_i = \pm f_o$:

$$\frac{f_o}{f_l}(m \pm 1) = n + m \tag{A-19}$$

Rearrange Eq. (A-19)

$$\frac{f_o}{f_l} = \frac{n + m}{m \pm 1} \tag{A-20}$$

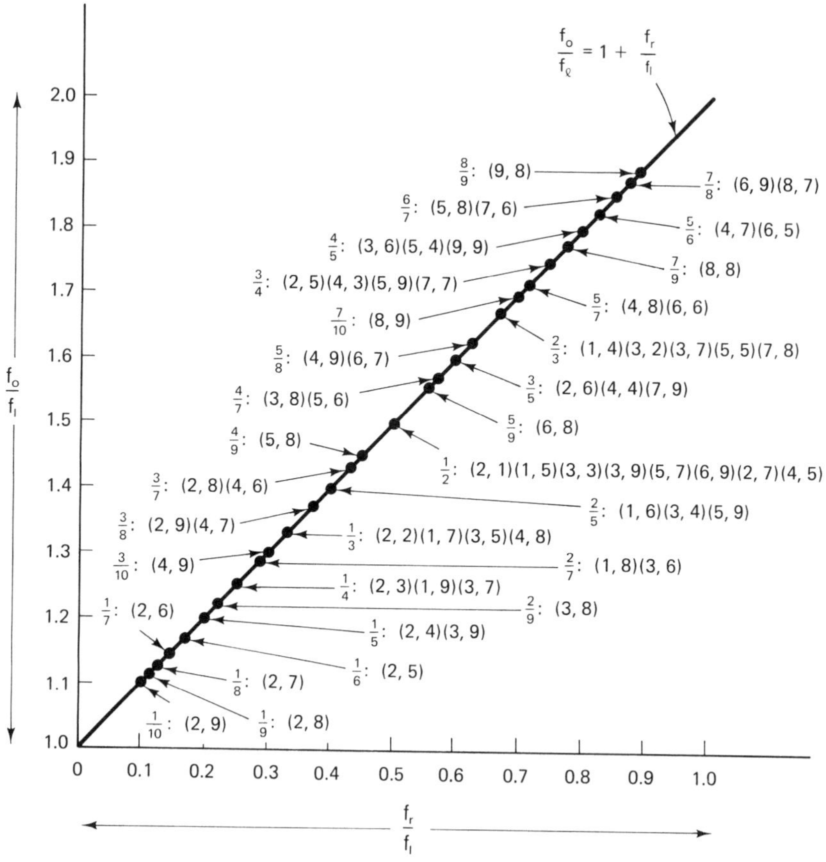

Figure A-3 Upconverter with output greater than either input

Substitute Eq. (A-20) into (A-13):

$$\frac{f_r}{f_l} = \frac{n + m}{m \pm 1} - 1 \tag{A-21}$$

Since f_r/f_l is between 0 and 1, the range of $(n + m)/(m \pm 1)$ must be between 1.0 and 2.0; therefore, the value of f_o/f_l is between 1.0 and 2.0.

Figure A-3 is of the function $f_o/f_l = f_r/f_l + 1$ and was derived from Eq. (A-13). The points on the lines are where interferences are present for integer values of n and m from 1 to 9. The notation for these points is the same as in Figs. A-1 and A-2.

Inspection of the points in Fig. A-3 reveals that the ratio of f_r/f_l should be selected for greater than 0.85 or less than 0.1 to obtain combinations of n and m where harmonically generated spurious power levels are considerably lower than the output f_o.

Index

A

B

C

D

F

N

O

P

R

S

T

U

V

Y